Historic American Roads

ON THIS MAP ARE SHOWN THE LOCATIONS
OF SOME OF THE PRINCIPAL HISTORIC ROUTES

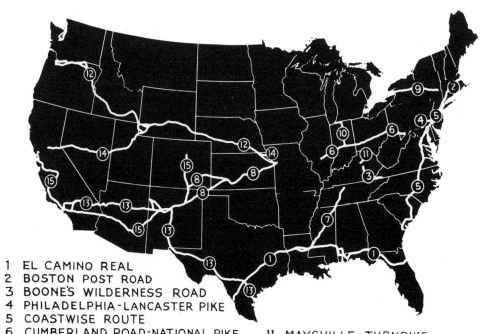

1 EL CAMINO REAL
2 BOSTON POST ROAD
3 BOONE'S WILDERNESS ROAD
4 PHILADELPHIA-LANCASTER PIKE
5 COASTWISE ROUTE
6 CUMBERLAND ROAD-NATIONAL PIKE
7 NATCHEZ TRACE
8 SANTA FÉ TRAIL
9 ERIE CANAL
10 MICHIGAN ROAD

11 MAYSVILLE TURNPIKE
12 OREGON TRAIL
13 CAMEL EXPRESS TO CALIFORNIA
14 PONY EXPRESS ROUTE
15 FIRST CROSS-COUNTRY TOUR

Historic American Roads

FROM FRONTIER TRAILS TO SUPERHIGHWAYS

Albert C. Rose
Illustrated by Carl Rakeman

CROWN PUBLISHERS, INC., NEW YORK

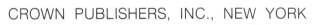

© 1976 by Crown Publishers, Inc.

Printed in the United States of America.

Published simultaneously in Canada by General Publishing Company Limited.

Paintings photographed by Taylor & Dull, Inc., New York

DESIGN: DEBORAH DALY

Library of Congress Catalog Card Number: 76-6224
ISBN: 0-517-525232
ISBN: 0-517-525496 pbk

CONTENTS

PROLOGUE

This pictorial history of public roads in Colonial America and the United States visually depicts the amazing development of highways in this country.

When the United States came into existence, it had a primitive transportation system. The rivers and the sheltered coastal waters, such as Long Island Sound, Chesapeake Bay, and Albemarle Sound, in the East, were the principal arteries for travel and commerce. Extending from these arteries were roads in various stages of development. A very few of these, near the largest cities, were "artificial roads," ditched and sometimes hard-surfaced with gravel or "pounded stone." The rest were improved only to the extent that stumps and boulders were removed and the worst irregularities of the ground leveled. Many of these roads in winter or during the spring thaw were impassable for wheeled vehicles. Bridges were few and far between. Travelers crossed small streams by fording, and larger ones by ferries. On the fringes of settlement, the "roads" were really only horsepaths for wheeled vehicles. But, by 1750, the roads were improved enough to establish a regular stage-wagon service from Philadelphia to New York via Trenton and Brunswick.

The first engineered road built in the United States was the privately constructed toll turnpike from Philadelphia to Lancaster, Pennsylvania. Built between 1793 and 1794 at a cost of $465,000, it was 62 miles long and surfaced with broken stone and gravel.

With the advent of the railroad era around 1830, road building in the United States virtually stopped for several decades. Interestingly enough, the stimulus for improved roads in the 1880s and 1890s came from bicycle riders, who launched a concerted campaign for better roads.

In 1893 an Office of Road Inquiry was established within the U.S. Department of Agriculture and charged with determining how to improve the national roads. This was the forerunner of the Bureau of Public Roads, which became today's Federal Highway Administration.

In 1907 the first inventory of national roads was published. It showed that of the more than two million miles of public roads, only 153,662 miles had any kind of surfacing. By comparison, there were then 213,904 miles of steam railroads in the United States.

In 1916, with the increasing popularity of the automobile, Congress passed the first Federal-Aid Highway Act, which created the long-standing federal-state partnership in roadbuilding and established the formula for distributing federal-aid funds to the states.

During the twenties and thirties most of the roadbuilding efforts in the United States were aimed at providing passable farm-to-market roads. In the late thirties President Franklin D. Roosevelt conceived the idea of a national, or interstate, system of highways. The 1956 Federal-Aid Highway Act, which created the Highway Trust Fund, finally made it possible for the program to become a reality.

An opportunity to present this overall highway history to the American public came in 1926, when the Bureau of Public Roads designed a historical exhibit as part of the federal government's contribution to the Sesquicentennial Exposition in Philadelphia. In planning this project, it became apparent that a reservoir of historical highway data would be an invaluable source from which to draw subject matter for future displays. The data were compiled by Albert C. Rose, a highway engineer with the bureau, over a period of more than twenty-six years. As a result, thirty-five dioramas, entitled "Highways of History," were constructed to illustrate scenes of public roads and related incidents in Colonial America and the United States. The sketching and painting of the dioramas themselves were done by the Bureau of Public Roads artist, Carl Rakeman. Oil-painted copies of the complete dioramas, produced by Rakeman, were subsequently shown at the New York World's Fair in 1940. Also, "Highways of History," a booklet containing reproductions of the dioramas and appropriate legends, was printed for general distribution. Ever since, there has been a widespread demand for this publication from the general public, state highway departments, highway and motor vehicle associations, and schools in this and other countries.

To make this historical data accessible to an even wider circle of highway interests, an expanded series of historical articles was begun. Each described a famous pioneer road, such as the Oregon Trail or the Santa Fe Trail, or a significant achievement and event, such as The First American Bridge, The Steam Road Roller, Rural Concrete Roads, the Inter-American and Alaska Highways, or the Interstate System. Publication began in the January 1950 issue of *American Highways,* the official magazine of the American Association of State Highway and Transportation Officials (AASHTO). Each article was illustrated with an original painting by Carl Rakeman and text by Albert C. Rose and maps by Margaret H. Davies. AASHTO then published the series in book form, using black and white illustrations. This work on the development of highways uniquely illustrates the astonishing growth of the public-road system in this country, which has no peer—either in extent or quality—in any nation, ancient or modern.

To present this information to the public in this Bicentennial period, the Federal Highway Administration is pleased to make Rakeman's now famous paintings, along with Rose's text, available in a new and expanded edition. We are particularly gratified that these paintings, which capture so well the drama of road building in America, can be seen in full-color reproductions for the first time.

Norbert T. Tiemann

January, 1976
Administrator, Federal Highway Administration

THE OLD ROAD BUILDER

For years articles describing the early history of road building, obviously based on the most careful and painstaking research, appeared in the magazine *American Highways*. The only indication of authorship had been that they were by The Old Road Builder.

It has since been disclosed that the author was Albert C. Rose, highway engineer and historian of the U.S. Bureau of Public Roads. The title he assumed is entirely appropriate. He was truly an old road builder, with experience that began in the mule-and-scraper days. Long experience coupled with a keen interest in highway development from the beginning of civilization made him an ideal historian.

Mr. Rose was born October 14, 1887, in Washington, D.C., of American parents. His parents moved to Philadelphia where he grew up. He graduated from Germantown Academy in 1905. For two years he sought to enter the Naval Academy at Annapolis, taking a preparatory course and passing entrance examinations, but was not admitted because the principal to whom he was alternate obtained the appointment. A second attempt fell short of success only because it was found that he would be overage at the time of entrance.

Discouraged with the East he decided to try the West. He arrived in Seattle in 1907 only a few weeks ahead of the financial panic of that year. He obtained a job as wireman with the Bell Telephone Company but lost the job almost immediately when the company laid off all but the oldest workers.

As a newcomer in the region, finding ways to eat became a problem. From November, 1907, to September, 1908, he worked at odd jobs such as electrician's helper, axeman for a surveyor, laborer on street surfacing (Vancouver, B.C.), laborer on construction of the Canadian Pacific Railroad, and laborer and tallyman in a lumbering camp.

Reversal in his fortunes came with the job as tallyman, from which he accumulated enough funds to enter the engineering school of the University of Washington in 1908.

He completed two years of the college course, spending his summer vacation with a field party locating the Columbia River Highway, which was to become famous as one of the first great scenic highway routes.

At the end of his sophomore year he had to leave college for a year to earn funds for further study. He became resident engineer on the construction of 17 miles of highway extending east from Spokane and costing $110,000. This was a big job for that period and one to tax the ability of any sophomore.

In October, 1911, he returned to college and completed his junior year. This ended his college career, but in 1924 the university awarded him a bachelor's degree as a result of his accomplishments. On leaving college he worked as computer, draftsman, and engineer for various firms in the Seattle area.

In 1913 he married Ella Lancaster, daughter of Samuel Lancaster, a leader in early highway development in the Northwest who played an important part in the construction of the Columbia River Highway.

From 1914 to 1919 Mr. Rose was roadmaster of Clatsop County, Oregon. During his tenure he launched the county on an active program of road improvement, demonstrating the advantages of engineering supervision and introducing competitive bidding for contracts in an area where negotiated contracts with attendant evils had been the rule. His three children, one boy and two girls, were born during this period.

When steps were taken in 1917 toward launching the program of federal aid to the states Mr. Rose was attracted by the wider opportunities he foresaw in a large national highway program. He applied to the Bureau of Public Roads for a position and took steps to qualify under Civil Service regulations. He received his appointment as highway engineer early in 1919 and was assigned to the Portland office of the bureau.

His first duties consisted of cooperation with state engineers in launching the federal-aid program in Oregon.

Largely on his own initiative he undertook a study of the causes of pavement failures. He soon discovered a close relation between the prevalence of failures and the character of the underlying soil. He then began a study of soils and the development of tests to measure characteristics. His tests, made in the field with simple equipment, are a landmark in the development of soil science as applied to highways.

As a result of this work, Mr. Rose was drafted in 1925 for the headquarters office of the bureau. There was some thought that he might continue his research on soils but he elected to go into the field of publications, exhibits, and motion pictures. He was soon placed in charge of work grouped under the general head of Visual Education. With increasing frequency it became necessary for him to conduct and supervise research on the early history of road building and use.

The history of early roads, their relation to the progress of civilization and their impact on the life, habits, and economy, fascinated him. Soon he was devoting all the time that could be spared from other duties to the subject.

Using the exceptional library facilities available in Washington, particularly those of the Library of Congress and the books and records on historic highways of the Bureau of Public Roads, he became an outstanding authority on highway history.

The Old Road Builder series is one of the important results of this work. He collaborated with Carl Rakeman, an artist of exceptional ability, in preparing a series of 109 paintings depicting highway scenes from the building of trails by the earliest settlers to the construction of modern express highways. Each painting was preceded by exhaustive research and they are believed to be authentic in every detail. Exhibits prepared under the direction of Mr. Rose, many of them dealing with historic matters, have been presented at state, national, and international fairs.

He assembled the results of many years of research and study in a book that will not only record the events of our highway history but will also interpret their effect on the development of the country and the lives and prosperity of its people.

It is believed that future generations will place Mr. Rose's name foremost among our highway historians. Following his retirement from the Bureau of Public Roads in 1950, he died at Holmes Beach, Florida, on June 24, 1966.

✻   ✻   ✻

Carl Rakeman, a native Washingtonian (D.C.), was educated at the Corcoran Art School and art academies in Dusseldorf, Munich, and Paris.

His art expression was not confined to any one medium. He was an etcher and a painter in watercolors, oils, and frescoes. He also worked in the field of mural decoration.

In 1921, Rakeman joined the Department of Agriculture, which at that time housed the Bureau of Public Roads—predecessor of the present Federal Highway Administration. During his many years with BPR, he painted exhibits for the Good Roads meetings, state fairs, and expositions such as the Brazilian Exposition (1922), the Century of Progress in Chicago (1933), an Overseas Exposition in Paris, the Golden Gate Exposition in San Francisco (1939) and the New York World's Fair (1940). In addition, he completed the series of 109 paintings depicting historic American roads, trails, and highways.

Rakeman early in his career was singled out by E. F. Andrews, founder and first director of the Corcoran School of Art (and an established artist in the District of Columbia), as the most qualified artist to copy the White House portraits of President and Mrs. Hayes, referring to him as "a young man of great talent." It was his first commission.

Important Rakeman works can be seen in Washington, D.C., and elsewhere throughout the country, such as in the U.S. Soldiers Home (Tennessee), the Ohio State House (Columbus), and the Hayes Memorial Museum (President Rutherford B. Hayes's home), Fremont, Ohio. He is extremely well represented at the latter, having worked continually for Colonel and Mrs. Hayes, painting numerous portraits of the entire Hayes family, and also restoring fine old paintings already hanging in the former President's home.

Among his most notable works are his murals in Washington, particularly the one adjoining the Senate Committee Room on Appropriations in the Capitol. The ceiling was frescoed by Constantino Brumidi, the highly regarded Italian-American artist. However, the lunettes over the windows, doorway, and fireplace were not decorated, and this work was placed in the hands of Carl Rakeman.

In each lunette Rakeman placed the portrait of a famous American general: George Washington, Anthony Wayne, Joseph Warren, and Horatio Gates. He framed each with an oval laurel wreath. Flanking each of the portraits are flags of the Colonial period, draped over contemporary helmets and arms. It is the only design in the Capitol showing the Colonial flags.

Rakeman retired from the Bureau of Public Roads in 1952, and died at the age of 87, in 1965, at Fremont, Ohio.

HISTORIC ROADS

1539-THE COMING OF THE HORSE

1539 THE COMING OF THE HORSE

Horses had been unknown in the New World since prehistoric times until the Spaniards introduced the forefathers of the modern animal. It was evident that the American Indians were unfamiliar with such a quadruped because they were stricken with terror or overcome with awe when the Spaniards first landed their mounts.

The cause of the disappearance of the prehistoric equestrian family in America may remain forever a mystery. Only fossils bear mute witness to their one-time existence. Perhaps they were exterminated by some sudden change of climate which produced a prolonged drought and the drying of all sources of drinking water. Possibly new-born foals were destroyed by vicious insect pests, such as the Hippobosca or the Ocstrus, which attack the umbilical cord and produce fatal ulcers unless thwarted by human intervention. Perhaps the incessant attacks of the increasing population of carniverous and predatory animals might have brought about the extinction of the wild horse. Annihilation by an epidemic is another alternative which has been suggested. Whatever cause, however, which erased the genus Equus from the ancient American scene, scientists are agreed that the Spaniards brought the ancestors of the present-day horse to the New World.

Christopher Columbus transported the first horses from Europe and disembarked them at Hispaniola (San Domingo) on his second voyage in 1493. Hernando De Soto, however, is believed to have landed the first horses which survived to renew the species upon the soil now occupied by the U.S.

De Soto's fleet of caravels entered a sheltered body of water which he named the bay of the "Espiritu Santo"(Holy Ghost) and which we call today Tampa Bay. Perhaps on May 28, 1539, as shown in the accompanying illustration, De Soto's men began landing more than 200 horses, westerly of the present city of Bradenton, Florida, where now stands a memorial marker erected by the National Society of the Colonial Dames of America.

When De Soto died, his followers consigned his weighted body to the Mississippi River in order to conceal the loss of their awe-inspiring leader from the Indians. Luis De Moscoso assumed command and directed the trip down the river."They shipped twenty-two of the best horses that were in camp: the rest they made dried flesh of." Harassed by the natives as they floated downstream, Moscoso decided to lighten the canoes by killing the horses."As soon as they saw a place convenient for it, they went thither and killed the horses and brought the flesh of them to dry it aboard. Four or five of them remained on the shore alive: the Indians went into them after the Spaniards were embarked – the horses were not acquainted with them and began to neigh and run up and down in such sort that the Indians, for fear of them, leaped into the water, and getting into their canoes went after the brigandines, shooting cruelly at them."

These half dozen horses, abandoned on the west bank of the Mississippi River by De Soto's men, perhaps are the ancestors of the feral or semi-wild horses of our western plains.

1540 - CORONADO IN NEW MEXICO

1540   CORONADO IN NEW MEXICO

The Spanish conquistador Francisco Vasquez de Coronado, led one of the most daring feats of exploration recorded in the annals of the New World. While serving as a regidor, or member of the Mexico City Council, Coronado was elevated by Viceroy Don Antonio de Mendoza, in 1539, to the Governorship of the Province of New Galicia, in northwestern Mexico, where was situated the northernmost Spanish settlement of San Miguel de Culiacan. Three years before, in April, 1536, Alvar Nuñez, Cabeza de Vaca, had straggled into Culiacan with three exhausted companions including a negro slave named Estevanico. These men were the sole survivors of the ill-fated expedition of Panfilo de Narvaez which landed near Tampa Bay, Florida, on April 15, 1528. After eight years of wandering and suffering as slaves of the Indians, Narvaez's party completed the first crossing of the continent by white men. The extravagant stories told by Narvaez about rich cities north of their route excited the cupidity of the Spaniards and initiated a series of northern explorations by land and sea. The probability of treasure in the north, like the gold which Pizarro had plundered from the Incas in the south, now eclipsed in importance the search for the mythical straits of Anian, or Northwest Passage, reputed to connect the Atlantic and Pacific Oceans.

With the dual objective of conquest and exploration Viceroy Mendoza instructed Governor Coronado to despatch Franciscan Friar Marcos de Niza upon a reconnaissance expedition to confirm the reports of Cabeza de Vaca. Guided by the negro Estevanico, the explorers, after a series of encounters with the natives, came into view of what appeared to be a province more wealthy than Mexico. Prevented by hostile Indians from making close inspection, de Niza hurried back to recount his supposed discovery of the Seven Cities of Cibola with their streets paved with gold.

Enthused by the friar's fanciful account, Viceroy Mendoza directed Governor Coronado to organize an expeditionary force with himself as Captain-General. The army marched out of Compostela in 1540. Plodding northward, Coronado about July 10, came in sight of and conquered the first of the cities of Cibola. The dissillusioned conquistadores, however, found that Friar Niza had mistaken golden streets for the sunlight reflected from the 5-stories-high adobe houses of the pueblo (village) of the Zuni Indians in the present western New Mexico. Stifling his chagrin Coronado continued northeastward past Inscription Rock and the existing pueblo of Acoma. His objective was the treasure cities in the northern Province of Quivira. Meanwhile an exploring party left the main army about August 25 and discovered the Grand Canyon of the Colorado River. The high-water mark of Coronado's exploration probably reached 100 miles north and east of Great Bend, Kansas. On the return trip, the conquistador followed closely the Cimarron River cut-off of the later Santa Fé Trail. The bedraggled remnant of his once colorful army returned to Mexico in the spring of 1542.

Inscription Rock, now El Morro (Spanish, headland or bluff) National Monument, shown in the illustration, is situated 35 miles east of Zuni pueblo and about 15 miles west of the Continental Divide.

1565~ SAINT AUGUSTINE

Saint Augustine, on the northeast coast of Florida, was the first permanent white settlement within the present territorial limits of the United States. This settlement became the parent hub from which radiated trails and roads to grow into the vast labyrinth of more than three million miles of highways now spreading from coast to coast. The first road built by white men in Florida and probably in our portion of North America joined the original crude wooden fort at Saint Augustine with Fort Caroline, (San Mateo) some forty miles to the north, situated on the St. John's River at St. John's Bluff about 17 miles northeasterly of the present Jacksonville and 1.5 miles east of Fulton.

Saint Augustine was founded by the Spanish Adelantado (Governor) Menéndez de Avilés who had been appointed by King Philip II of Spain to crush the French Huguenot colony at Fort Caroline considered a menace to the Spanish trade route. Arriving off the coast of Florida on August 28, 1565, Menéndez named Saint Augustine in honor of that date dedicated to the festival of the eminent saint of his church. A few years previously, Jean Ribault had made an unsuccessful attempt to establish a French settlement at Port Royal, South Carolina, in 1562, by permission of King Charles IX of France. Admiral Coligny, the French Huguenot leader, despatched another expedition, in 1564, under the command of Réné de Laudonniére who landed in the harbor of Saint Augustine which he named the River of Dolphins because of the abundance of dolphins (porpoises) swimming at the mouth of the stream. Later, Laudonniére coasted north and examined the St. John's River (called then the River May and later River San Mateo - Spanish for Saint Matthew) before building a fort about two leagues (5.3 miles) from the mouth of the stream, upon a pleasing hill of "mean height" in the midst of the villages of the Timucuams.

Jean Ribault arrived with shiploads of reinforcements and supplies. The news of this expedition motivated King Philip II of Spain to organize a fleet in command of the brave and remorseless soldier, Pedro Menéndez de Avilés, with orders to drive the French from Florida. Plagued by storms and accidents about one half the 11 vessels, with 2,600 men, which departed from Spain, cast anchor in the harbor of Saint Augustine on August 28, 1565. Menéndez with a force of 500 men, including 300 arquebusmen, and the remainder pikesmen and targeteers, marched northward to attack the French at Fort Caroline. Twenty Asturians and Basques, under their Captain, Martin de Ochoa, blazed with axes a path through the forests and swamps guided by a compass in the hands of Menéndez. Attacking at the break of dawn on September 20, more than half the Frenchmen in the fort were killed, fifty women and children were captured and the remaining defenders escaping with the French commander, Laudonniére, were rescued.

The southern portion of the trail opened by Menéndez between Fort Saint Augustine (later Fort San Marcos, now Fort Marion) and Fort Caroline is said to coincide with the present United States Route 1 which follows the King's Road, opened in 1765.

3

1607 – THE INDIAN CANOE

The canoe was the principal vehicle of the Indian aborigines when the Englishmen landed in the New World at Jamestown, Virginia, in 1607. The colonists used the Chesapeake Bay, the James and other rivers as water highways leading into the interior. Dense forests of oak, walnut, elm, chestnut, cherry, mulberry and other trees covered the hills and valleys down to the sandy beach bordering the waves of the sea. Through this wilderness, over a labyrinth of trails, wild animals such as the deer, wolf, bear, fox, opossum, polecat, weasel, mink and many others found their way to water, food and salt licks. The moccasined feet of Indians pattered softly over former animal paths, or along trails opened by their own traffic, on errands of barter between tribes, hunting and fishing, and tribal warfare. With their bows and arrows the Indians brought to earth birds for food including the wild turkey, quail, partridge, pigeon, goose and duck. From the silver streams they fished the sturgeon, trout, mullet, perch and eel. From the bay they gathered the oyster, crab, shrimp and muscle. In all these activities the favorite vehicle of the Indians was the canoe. There were no horses in Virginia until 1609, two years after the colonists first landed. The Virginia Indians did not possess any of the horses introduced by the Spaniards. The dusky forest denizens for centuries had employed rude hand litters to move their sick. Dead as well as live loads were transported from place to place upon the broad backs of stalwart aborigines. Their faithful dogs were the only burden bearers the savages knew.

In the illustration, the redoubtable Captain John Smith is depicted in conference with the proud and powerful Indian chief, Powhatan, at his main village situated upon the north bank of the James River about one mile downstream from the site of the present city of Richmond, Virginia. Crossing the log bridge at the right is Pocahontas, the favorite daughter of Powhatan. Her name is written indelibly into the history of the Old Dominion because of the romantic manner in which she later saved Captain John Smith's life. The Indians at the left are building a dugout.

The dugout was the type of canoe favored by the southern Indians in smooth waters. To quote from Captain Smith's writings, "Their fishing is much in boats. These they make of one tree by burning and scratching away the coals with stones and shells, till they have made it in the form of a trough. Some of them are an ell (English = 45 inches) deep, and 40 or 50 feet in length, and some will bear 40 men, but the most ordinary are smaller, and will bear 10, 20, and 30, according to their bigness. Instead of oars, they use paddles, and sticks, with which they will row faster than our barges."

1611
THE FIRST AMERICAN BRIDGE

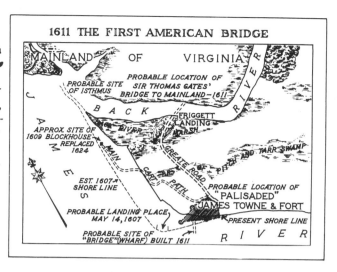

1611 THE FIRST AMERICAN BRIDGE

The first "bridge" on record built by the English settlers in America was located at "James Towne" island, Virginia, where the actual landing was made on May 14, 1607. This structure was not a stream crossing in the sense that the term is customarily used, but a wharf about 200 feet in length extending from the James River bank to the nearby channel where the 12-foot depth of water provided docking facilities for the small caravel ships. This so-called "bridge" was built in 1611, according to a letter written by Sir Thomas Dale, on May 25, 1611, "To the President and Counsell of the Companie of Adventurers and Planters in Virginia" in which Dale stated that immediately upon his arrival /1611/ to succeed Lord de la Warre as Deputy Governor "Captain Newport undertook the Bridge with his Mariners," at Dale's direction, "to land our goods dry and safe upon." Another reference to this same "bridge" was made in "A Briefe Declaration of the Plantation of Virginia duringe the first Twelve Yeares," (probably dated about 1625).

It seems incredible that four years should elapse between the date of the landing of the first settlers and the time construction began on a wharf approach to the island or on a bridge connecting the island with the mainland for a number of reasons: (1) Captain John Smith records the character of the bridges built by the Indians on the Virginia mainland when "coming ashore, landed amongst a many of creeks, over which they were to pass such poor bridges, only made of a few cratches /tree trunks with forks/ thrust in the ose /muddy ooze/, and three or four poles laid on them, and at the end of them the like, tied together only with barks of trees, that it made them much suspect those bridges were but traps, which caused Smith to make divers savages go over first, keeping some of the chiefs as hostages until half his men were passed to make a guard for himself and the rest;" (2) among the names of the first planters recorded June 15, 1607, there were the four English carpenters, William Laxon, Edward Pifing, Thomas Emry and Robert Small; (3) as illustrated in "The Generall Historie of Virginia, New England, and the Summer Isles with the names of the Adventurers, Planters, and Governoors from their first beginning An. 1584 to the present 1624" there were three wooden trestle bridges built in 1620 on the Bermuda Islands, showing their familiarity with this type of structure.

Furthermore, for the foregoing reasons, even though the source material cannot be traced, it seems highly probable that Philip Alexander Bruce was correct when he stated in his "Economic History of Virginia in the Seventeenth Century" that under the energetic direction of Sir Thomas Gates when he returned to Virginia, in 1611, "A bridge was built to connect the island with the mainland" as shown in the accompanying illustration. The location of this bridge is indicated on the sketch as probably at the terminus of "the old Great Road" at the "Friggett Landing" on Back River where supplies were delivered to the settlement. No ancient charts of the island for the period of 1607-1698 have been discovered.

1612–THE GREAT SAUK TRAIL

The Great Sauk Trail, or Potawatomi Trail, was an overland variant, across the present State of Michigan, of the southernmost of the two main long-distance routes, shown in the map (right), which the French had discovered between New France (Canada) and the Mississippi River valley. Over the northern branch Etienne Brulé probably was dispatched by Champlain to Georgian Bay in 1612. The Great Sauk Trail was named for the Sac Indians who with the Fox Indians used this path in their travels from eastern Canada to the far northwest. The Iroquois of New York followed the Great Sauk Trail in their warfare with the Miamis, Illinois and other western tribes. The Potawatomi village of Pokagon, named for its wise chief, was situated on the west side of the St. Joseph River a little south of the trail.

The Great Sauk Trail branched from the southern transcontinental trail at the "place of the strait" (French, place du detroit), the Detroit of today, founded by the Frenchman Antoine de la Mothe Cadillac on July 24, 1701. The Great Sauk Trail ran westerly to the portage between the Checagou (Chicago) and Illinois Rivers considered the strategic key to the American continent, and continued to the existing Green Bay, Wisconsin, region – the habitat of the Sac and Fox Indians.

Over the Great Sauk Trail ebbed and flowed the tides of four great eras in the economy of the vast Northwest Territory bordering upon Lake Michigan: (1) The period of the French exploration from 1630 to 1680; (2) the times of the fur traders, comprising three nationalities, the French, the English and the Americans, which reached a crest about 1825; (3) the years when the emigrants swarmed to the Northwest and earned their livelihood as farmers, lumbermen, miners and in allied occupations from 1826 to, say, 1900; and (4) the industrial age, considered to have begun with the formation of the initial units of the immense automobile industry, dating from 1901 to the present time.

During the second epoch, General Anthony Wayne defeated the Northwest Indians at the Battle of Fallen Timbers (near Toledo, Ohio) and, in 1795, at the ensuing Treaty of Greenville, the American Government was ceded, "a piece of Land Six Miles Square at the mouth of the Chicago River." Here, in 1803, Captain John Whistler, grandfather of the celebrated artist, erected the stockade, shown in the accompanying illustration, named Fort Dearborn in honor of Secretary Henry Dearborn in the cabinet of President Thomas Jefferson.

Colonel (later General) Lewis Cass, military governor of Detroit during the War of 1812 with England, was appointed Governor of Michigan Territory by General William Henry Harrison in 1813. Aware of the military handicaps resulting from the lack of roads into the interior, Governor Cass, after peace was concluded, proposed that the War Department should build highways radiating from Detroit. A responsive Congress authorized five highways, one of which was to traverse the Great Sauk Trail. Construction was begun on the Chicago Road in 1827. Stage coaches were running over the entire 295-mile distance from Detroit to Chicago by 1833.

1625~PAVED STREETS IN MAINE

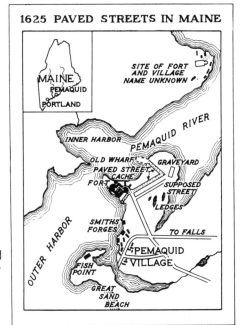

1625 PAVED STREETS IN MAINE

Some of the streets were paved at Pemaquid, Bristol Township, Maine, probably before the year 1625. Although the original settlement antedates historical record some authorities believe that Pemaquid was coeval with the Plymouth colony founded in 1620. Other historians assign to Pemaquid a more ancient lineage.

In pre-Colonial times Pemaquid was the center of the most prolific fishing grounds along the coast of North America. Europeans were attracted first to this region because of the priceless marine life. Thornton wrote, "To Pemaquid we must look for the initiation of civilization in New England." Pemaquid was visited by David Ingram as early as 1569, by Captain Bartholomew Gosnold in 1602, by Raleigh Gilbert in May 1607, and by Captain Thomas Dermer in 1619. The short sand beach at Pemaquid, in contrast to the generally rock-bound coast of New England, provided a convenient landing for fishing boats. The Sieur de Monts while exploring the coast in 1605, with the Frenchman Samuel de Champlain, observed settlements in this vicinity. Captain George Weymouth of England also explored this region in 1605. When the English Captain John Smith visited the island of Monhegan, in 1614, he saw shoreward "in the Maine" at the port of Pemaquid a ship owned by Sir Francis Popham whose associates had cast anchor there for "many years."

It is possible that merchants with headquarters in Bristol, England, or perhaps London, maintained a fish and fur trading center at Pemaquid as early as 1600. Two decades later, in 1622, there were thirty ships engaged in the fish and fur trade in the Pemaquid area taking advantage of the excellent harbor and the abundant fish bait at the falls of the Pemaquid River.

Among the many artifacts which have been unearthed in the neighborhood of ancient Pemaquid probably the street paving is the relic which more than any other establishes the advanced state of civilization at this European outpost. The identity of the builders of this paving has defied the painstaking researches of historians and archaeologists. John Henry Cartland, who is an authority upon the subject, in his "Ten Years at Pemaquid" describes the construction of "what appears to be a short section of a street about ten feet above high water mark, leading down a fine easy sloping field toward a small beach,*****," as shown on the map above.

According to Cartland, "The larger stones form what we term the main street, which is thirty-three feet in width including the gutters, or water courses. The finer work of cobble-stones evidently taken from the beach nearby is eleven and one-half feet wide. The longer cobbles were selected and placed across the sidewalk on lines two feet and one-half apart, then the space filled in with smaller ones."

Finally, according to a report sponsored by the Maine Historical Society and written by its Secretary Edward Ballard, dated August 25 and 26, 1869, the opinion was advanced, ***** The regular arrangement of the beach-stones, the depression for the water course to the shore, the curbstones, the adjoining foundation-stones still in place, ***** proved, beyond the possibility of a doubt, that a European community had dwelt on this spot."

1632–FIRST HIGHWAY LAW

The first highway legislation in the British Colonies in North America was passed by the Legislature of Virginia, possibly meeting, in September 1632, in the third wooden church, shown in the accompanying illustration, the eighth year of the reign of King Charles I of England. ACT L provided that, "HIGHWAYES shall be layd out in such convenient places as are requisite accordinge as the Governor and Counsell or the commissioners for the monthlie corts shall appoynt, or accordinge as the parishioners of every parish shall agree."

Twenty-five years later this basic highway law of Virginia was supplemented by the Legislature in March 1657–8, in the ninth year of the Commonwealth. Act IX, "Concerning Surveyors of High Waise" provided, "That surveyors of highwaise and maintenance for bridges be yearly kept and appointed in each countie court respectively, and that all gennerall wayes from county to county and all churchwaies to be laied out and cleered yearly as each county court shall think fitt, needful and convenient, respect being had to the course used in England to that end."

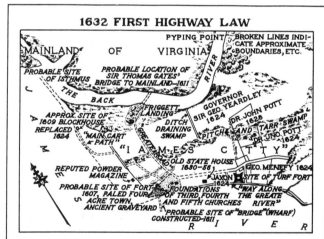

In the mother country of England during the feudal times of the Middle Ages the care of the roads was based upon the trinoda neccessitas (the threefold required service) of the tenant, namely: the duty of repelling an enemy, the construction of fortifications and the repair of roads and bridges. With the decline of feudalism, the foregoing requirement developed into the common law whereby the repair of highways became the responsibility of the local parishes or hundreds which were traversed. Neglect called for the indictment of the guilty authorities by the Court of Quarter Sessions. Compulsory labor upon the highways was legalized, in 1555, when the "Statute for Mending of Highways" specified that, "Constables and Churchwardens of every parish shall yearly, upon the Tuesday or Wednesday in Easter week, call together a number of parishioners, and shall then elect and choose two honest persons of the parish to be surveyors and orderers for one year of the works for amendment of the highways in their parish."

The statute labor system was introduced to America by the first settlers and became embedded in the laws of the British Colonies and later the United States where they survived for three hundred years until the beginning of the twentieth century. The major portion of highway improvement was performed by statute labor both in the British Colonies and in the infant United States until the first turnpike was begun in Virginia in 1785. Following the pattern of the Commonwealth of Virginia and the English homeland, highway laws were enacted in Massachusetts, in 1639; in Connecticut, in 1643; in New York, in 1664; in Maryland, in 1666; in New Jersey, about 1675; and in South Carolina, in 1682.

The lack of legislation authorizing local taxes for roads combined with the iniquities of the statute labor system were responsible largely for the bad condition of the roads in the Colonies and early Republic.

1636~THE CONNECTICUT PATH

The Connecticut Path, the Old Connecticut Path, the Bay Path, the Old Bay Path, and the "New Way" were the names given by the Pilgrim Fathers to the Indian trails connecting the Massachusetts Bay colony with the first white settlement along the Connecticut River. The settlers learned of the existence of these trails from the Indians who brought corn from the Great (Connecticut) River valley to sell in Boston. One of the early Indian travelers was Wahginnacut, sachem of the Podunk tribe whose habitat extended along the Connecticut River as far north as Agawam (Springfield). The first white man to travel the Indian trail to the west was the prospector John Oldham, in 1633, who endorsed the Connecticut River Indians' invitation to the white men to settle in their valley. As a result John Cable and John Woodstock were sent forward over the Old Connecticut Path, in the spring of 1635, by William Pynchon, a gentleman from Springfield, in Essex, England, who had established his plantation on the rocks of Boston Neck, now called Roxbury.

The nomenclature of Connecticut Path and Bay Path was employed interchangeably by the Pilgrim Fathers with respect to the continuous Indian trail westward from the parent bay settlement across Massachusetts and Connecticut. For this reason uncertainty exists concerning the exact identity of these paths. The original Connecticut or Bay Path ran from New Town (Cambridge) through the present Weston, Wayland, Framingham, Ashland, Hopkinton, Westborough, Grafton, Oxford, Charlton, Sturbridge, Brimfield to Springfield. This original path became known as the Old Connecticut Path or the Old Bay Path when the "New Way" was opened from Weston through the present Sudbury Center, Marlboro, Worcester and Brookfield and joining the old path at Brimfield. Later this trail was continued from Springfield northwestward across the Berkshire Hills to Fort Orange (Albany) where junction was made with the great Iroquois or Mohawk trail meandering westward across the today's New York State. The Connecticut or Bay Path in Massachusetts blazed the route for the subsequent Upper Boston Post Road and later the Boston and Albany Railroad, while the Mohawk trail established the location of the New York Central Railroad.

The first white settlement at the Indian Suckiaug (Hartford, Connecticut) was sponsored by the Dutch from New Amsterdam, in 1633. The first English settlement was built, in 1635, by sixty emigrants from New Town (Cambridge, Massachusetts). The main emigration, however, which depleted the populations of Dorchester, Watertown and New Town, set out on May 31, 1636, under the leadership of Thomas Hooker, pastor of the church at New Town, and Samuel Stone. There were about one hundred people in the party. They drove with them one hundred and sixty cattle to supply milk along the way. Pastor Hooker's sick wife rode a horse litter, shown in the accompanying illustration. Little three-year-old Samuel Hooker shared the rough ride with his mother. They named their destination New Town after the village they had left. This name was retained for a year until changed in 1637, to Hartford in memory of Stone's birthplace in England. Because the Hooker party left no record of their route there has been much speculation about the precise path they followed.

1673~ COLONIAL POST RIDER

The Boston Post Road was the first route over which Colonial mails were carried between the settlements in New England and New York. The region traversed by this road was a dense wilderness penetrated by wild animal and Indian trails for more than half a century following the first permanent settlement at New Amsterdam by the Dutch, in 1613, and the landing of the Pilgrim fathers at Plymouth, Massachusetts, in 1620. The metropolitan centers of these settlements later shifted to the best harbors, one at Boston and the other at New York.

From these shipping points the colonists advanced the outposts of their civilization westward into the wilderness progressing steadily step by step by means of treaty and warfare with the Indians. In 1633, following the Bay Path, the emigrants from the Massachusetts colony settled in the valley of the Connecticut River at the site of Springfield. The Upper Boston Post Road follows the general location of this Bay Path. The direction of the later Upper and Middle Boston Post Roads in Connecticut was fixed by the settlement of Hartford, in 1635, by the English. The general

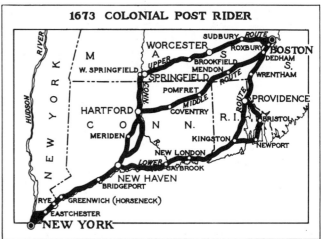

course of the Lower Boston Post Road through Rhode Island was established by the settlements of Roger Williams and his coreligionists, in 1636.

The initial step in favor of intercommunication from Boston to New York was taken on January 22, 1673, when the English Governor, Francis Lovelace, of New York, desirous of cementing relations with New England, despatched the first post rider toward the north by way of New Harlem, Williamsbridge, Eastchester, Horseneck (Greenwich), New Haven, Hartford, Springfield and Roxbury to Boston. This parent mail service in English-speaking America traversed the interior route which became known as the Upper Boston Post Road. The foregoing back-country settlements, between New York and Boston, bear mute witness to the rapid growth in population, the conquest of the wilderness, the declining fear of Indian attack, and the improvement of horse paths during a period of less than three generations.

The population of New England now totaled roughly 100,000 resolute pioneers housed in comparative safety. Only nineteen years before, in 1654, the residents of New Amsterdam had erected a wooden palisade across lower Manhattan Island as a protection against Indian attack, as shown in the accompanying illustration. It was situated along the north side and gave its name to the present Wall Street. Between the two main settlements ferries at stream crossings were installed at an early date. The one at New London was always a source of delay and inconvenience because of the width of the river.

The total length of the Upper Boston Post Road in 1800, was 250 miles; the middle route was 203 miles long in 1814; and the lower road from 247 to 259 miles in length depending upon whether the east or west branch was selected beside Narragansett Bay.

1679—THE PORTAGE PATH

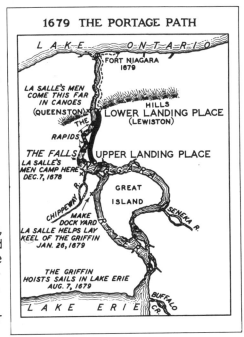

1679 THE PORTAGE PATH

From the time of their first settlements in New France (Canada) the French explorers searched for the Northwest Passage, the legendary Straits of Anian, as a short cut to the fabulous wealth of the Indies. As early as 1634, Jean Nicolet, an agent of the French fur trader, Samuel de Champlain, set out on a western tour of discovery aimed at finding the best route to China. Because the waterways provided paths by which the primeval wilderness could be traversed with the least effort and a minimum of peril, it was natural that the French explorers should penetrate the interior by way of the St. Lawrence River and the Great Lakes. In the course of their journeys it was necessary to bypass rapids and falls by carrying their canoes, supplies and equipment over parallel land paths. There were other transverse trails across ridges separating two water courses where resort had to be made to overland carriage. These paths beside non-navigable sections of a stream or between two navigable waters were known as portages. The word is derived from the French porter, meaning to carry. During the seventeenth century when the French explorations were at their peak these portages were the most important land routes. Admitted that the wilderness was criss-crossed with an extensive labyrinth of wild animal and Indian land trails, nevertheless, these paths were of secondary importance to Frenchmen who paddled across the rivers and lakes as their principal arteries of travel. Furthermore, of all the many portages in the New World, the one around Niagara Falls was of greatest strategic importance to the French. The total length of the Niagara River was 34 miles. It was navigable for 20 miles from Lake Erie to the upper rapids. There began the seven-mile portage around the gigantic cataract averaging a fall of 160 feet to the lower river at the site of the present Lewiston. Thence the canoes could be propelled again down the navigable river for the remaining seven miles to its debouchment into Lake Ontario.

In the illustration, may be seen René Robert Cavalier, Sieur de La Salle, the most famous of the French explorers and a few of his party, on January 22, 1679, on the portage around Niagara Falls. The celebrated adventurer is leaning upon the shoulder of the tonsured Recollect Father, Zénobe Membré, who is clad in a coarse gray capote, with the cord of St. Francis about his waist, a rosary and a crucifix hanging at his side, and his feet shod with sandals. At the right stoops a native common laborer, or engagé, carrying a pack balanced by a tump line encircling his forehead. Another engagé rests cross-legged upon his pack, while two of his comrades carry a canoe at the left.

1700 – THE IROQUOIS TRAIL

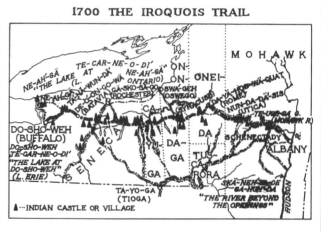

The Iroquois Trail, or Mohawk Trail, was the great central thoroughfare of the Iroquois Confederacy (Ho-de'-no-sau-nee = "the people of the long house") across the present State of New York, connecting the site of Albany on the Hudson River with the site of Buffalo beside Lake Erie. This ancient Indian Trail (Wä-ä-gwen'-ne-yuh) ran through the center of the Long House of the Five Nations (Six Nations after the accession of the Tuscaroras in 1712) beginning with the Mohawks on the east through the villages of the Oneidas, Onondagas, Cayugas, Senecas and Tuscaroras in the west. This Long House was the symbolic common shelter of the several nations. Like their long bark dwelling houses with separate apartments, with a central fire, each accommodating two families, the Iroquois Trail joined the castles or principal villages, of each nation where the council fires burned.

This beaten path was trodden by generation after generation of red men over its meandering original 360-odd-mile course beneath the overhanging trees of the forest. The trace, varying in width from twelve to eighteen inches, was a well-worn groove from three to twelve inches in depth depending upon the stability of the natural ground. It was the most direct route between the Hudson River and Lake Erie following the best topography with amazing precision. From the present Albany (originally Skä-neh'-ta-de = "beyond the openings") the trail proceeded northwest to today's Schenectady and there divided into two branches along both banks of the Mohawk River to reunite at the portage at Rome (Dayä-hoo-wä-quat = "place for carrying boats") about one mile long over the Atlantic Ocean-Great Lakes divide to Wood Creek leading to the Oswego River connection with Lake Ontario. From Rome the main trail extended westward, skirting the northern shores of the Finger Lakes and arrived at Batavia (De-o'-on-go-wä = "the grand hearing place") where the sound of the Tonowanda Creek rapids first was heard and where even the roar of distant Niagara Falls could be detected by the practiced ears of the Indians. Thence the trail came upon the site of Buffalo at the head of Main Street and descended Buffalo Creek to its western terminus on Lake Erie. This principal east-and-west path was the base of an inverted triangle of Iroquois trails which had its apex at Tioga on the Susquehanna River joining there the main trail leading south through Pennsylvania.

Swift-footed Iroquois runners carried messages for the entire length of the main trail from Albany to Buffalo in three days. The trained runner, shown in the illustration, could cover one hundred miles a day. When relays were resorted to the length of the day's journey could be increased considerably. During the day the trail was followed easily by the Indians. At night the runners were guided in the fall and winter by the constellation Pleiades (Got-gwär-där = "in the neck of Taurus – The Bull"), a group of seven stars visible to the naked eye in midwinter toward the east in the evening, overhead at midnight, and toward the west in the morning. In the spring and summer the runners were directed by a four-star group which they called Gwe-o-gä-oh (the Loon).

1751—THE PENNSYLVANIA ROAD

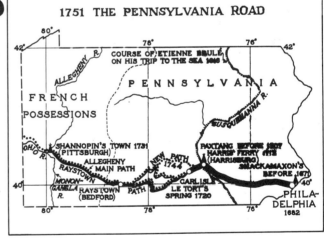

1751 THE PENNSYLVANIA ROAD

The Pennsylvania Road, throughout the greater portion of its length, now known as United States Route 30, was in pre-Revolutionary times the most important sunrise-to-sundown pathway connecting the colonies along the Atlantic seaboard with the vast hinterland beyond the summits of the Allegheny Mountains, lapped by the waters of the Great Lakes and the Ohio River and its tributaries. This great thoroughfare to the west was transformed within three centuries from an Indian trail and fur traders' path into one of the country's most important east-west motor vehicle arteries.

The Pennsylvania Road was the main trans-mountain route of the British Colonies in America because the Province of Pennsylvania was situated geographically so as to provide the shortest path over which our forefathers could pursue their primary objectives—a northwest short cut to China—while expanding their secondary objectives—homes in the wilderness and trade with the Indians and the mother country. The prehistoric Indian trail, known as the Allegheny Path, which slowly developed into the Pennsylvania Road, began at the Delaware Chief Shackamaxon's Indian village, on the site of the present city of Philadelphia, and rose and fell across the successive mountain passes and valleys to the junction of the Allegheny and Monongahela Rivers, and beyond.

Etienne Brulé, Champlain's interpreter in the years 1615–1616, was probably the first white man to cross this westbound trail in a north-south direction. The Indian trader, James Le Tort, journeyed to the Allegheny River over this path in 1727. Carlisle was the beginning of the pack-horse trail to the west, shown in the accompanying illustration, and the end of the wagon road from the east, until 1755, when Governor Morris of Pennsylvania ordered wood choppers to open a road through the forest to a point about 25 miles west of Raystown (Bedford) to serve as an auxiliary supply line for General Braddock in his disastrous campaign against Fort Duquesne (Pittsburgh).

The Pennsylvania Road, between Raystown and "Pitt's Borough", was called Forbes' Road because it was opened, in 1758, at the direction of General John Forbes, as the transportation and supply route for his victorious campaign against Fort Duquesne. It was at the outset of the Revolutionary War that the Continental War Office, located at Philadelphia, designated the then Forbes' Road, the main military route to the west, as the Pennsylvania Road. Similarly Braddock's Road, the principal thoroughfare from Virginia and Maryland was named the Virginia Road.

With the completion of the Pennsylvania Canal system to Pittsburgh, in 1834, the Pennsylvania Road and the National Pike lost much of their business. The opening of the Baltimore and Ohio Railroad to the Ohio River, in 1852, in turn eclipsed the Pennsylvania Canal. Outmoded by the railroad, the Pennsylvania Road, like all other wagon roads, fell into disuse and lack of repair until interest was revived in highways first by the League of American Wheelmen, about 1885, and later by the advent of the "horseless buggy" in the "Nineties."

1753—WASHINGTON CROSSING THE ALLEGHENY

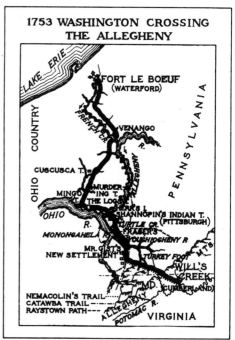

1753 WASHINGTON CROSSING
THE ALLEGHENY

In the wilderness far removed from the frontier settlements there were no canoe ferries at river crossings. Woodsmen improvised rude log rafts built from the forest trees. The logs were bound together with withes – slender, flexible branches of willow or osier. In the accompanying illustration Major George Washington is shown with Christopher Gist poling their hastily constructed craft across the Allegheny River in the dead of winter. This river crossing was the sequel to the action of a French army which had invaded the Allegheny River valley from Canada for the strategic purpose of gaining control of the Ohio River region so as to confine the British colonists to the narrow territory east of the Appalachian mountains. Immediately upon learning of the invasion Governor Robert Dinwiddie of Virginia decided to challenge the bold attempt to seize territory claimed by the British Crown. His first problem was to find a suitable courier to carry an ultimatum across hundreds of miles of forest wilderness intervening between the Virginia settlements and Fort Le Boeuf (Waterford, Erie County, Pennsylvania), the nearest French outpost. Governor Dinwiddie solved the problem by choosing the 21-year-old Virginian who later became "The Father of his Country."

Major Washington set out from Williamsburg, Virginia, on October 31, 1753. Proceeding to Will's Creek (Cumberland, Maryland), at the outskirts of the settlements, Washington employed as his guide the experienced woodsman Christopher Gist. With four other companions the pair left Will's Creek on November 14 and followed the Delaware Indian Nemacolin's trail as far as Gist's plantation. Thence they rode their horses over the Catawba Trail and the Raystown Path to the Delaware Chief Shannopin's Indian town (Pittsburgh, Pennsylvania) where there was a crossing of the Allegheny River.

Leaving on December 16, 1753, for the return trip, Washington and his party paddled down the river in a French canoe to Venango where they arrived on December 22. Here the French induced the Indian helpers to desert. Washington and his white companions pressed forward on horseback. After three days, progress was so slow that Washington decided to relinquish the horses and baggage to the custody of the interpreter, Van Braam. The resourceful Major struck out on foot with Christopher Gist by the most direct route homeward through the woods. Arriving at the northerly bank of the Allegheny River on December 29, they hastily assembled the crude raft of logs and poled across to the island (Herr's) in the river above Shannopin's Town. On the way over Washington fell into the icy waters but saved himself from drowning by clinging to the raft. They thawed out the Major's clothes and Gist's frozen fingers that night beside a camp fire lighted upon the island. The night was so cold that the surface of the river was frozen thick enough by the next morning to enable them to walk the remaining distance to Shannopin's Town across the ice. Thence the sturdy couriers tramped to John Fraser's place at the mouth of Turtle Creek. Major Washington returned to Williamsburg, Virginia, on January 16, 1754, and delivered to Governor Dinwiddie the refusal of the French commander to heed the warning.

1755—BRADDOCK'S ROAD

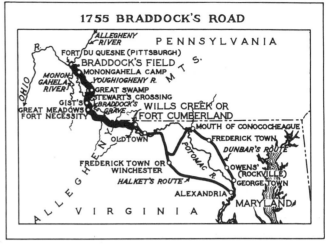

1755 BRADDOCK'S ROAD

General Edward Braddock's failure to capture Fort Du Quesne (Pittsburgh, Pennsylvania) spelled out in flaming letters a defeat which has been passed down from generation to generation of Americans as the crowning example of military folly. The basic trouble, however, was the lack of a good road. This strategic disaster followed in the wake of the unsuccessful attempt to perform the herculean task of cutting through a forest wilderness a new wagon path over which to march an army of veterans fresh from European battlefields and supplied with long trains of covered wagons. The route followed the trail used by Major George Washington and Christopher Gist in 1753, in their journey to warn the French commandant at Fort Le Boeuf. There were six towering ranges of the Allegheny Mountains that had to be crossed between Fort Cumberland, the British concentration point at Wills Creek, Maryland, and the French stronghold at Fort Du Quesne. General Braddock had been advised by his young military aid, Major George Washington, to march his troops in single file over the existing Indian trails and to transport his supplies upon the backs of pack animals. The British general, however, seasoned in the traditional military science of the Old World chose the method which seemed to promise the greatest concentration of striking power.

Major General Braddock arrived at Hampton Roads, Virginia, on January 14, 1755, with the authority of generalissimo of the British forces

A detachment of 600 soldiers and axmen, commanded by Major Russell Chapman, on May 30, began broadening the trail into a wagon road 12 feet wide from Fort Cumberland. Between June 7 and 10, the army began to march in three divisions. The progress made by the troops was painfully slow—from 2 to 5 miles a day. Mountains had to be scaled, streams bridged and morasses made passable. The road was so narrow that the wagons soon were strung out over a distance of four miles, just as Benjamin Franklin had predicted when he arranged to furnish General Braddock with Pennsylvania wagons. At that time the English officer's reaction had been that, "These savages may indeed be a formidable enemy to green raw American militia; but upon the King's regular and disciplined troops, sir, it is impossible they should make any impression." The general had yet to learn of the impenetrable obstacle placed in his path by a forest wilderness through which his troops struggled when weakened by sickness from a diet of salt-cured food and harrassed by prowling bands of enemy scouting parties. Thus encumbered, it took the army more than two months to traverse the 115 miles of mountain wilderness. At last, on July 9, the 1,200 soldiers constituting the advance guard were brought to attention in well-formed ranks in a level space beside the Monongahela River some ten miles from Fort Du Quesne. Suddenly rifles cracked as a murderous fire was delivered at the bewildered soldiers by eight hundred Indians and French ambushed in the surrounding woods. Two-thirds of the British forces fell killed or mortally wounded. General Braddock died four days later. The survivors broke and beat a wild retreat to Fort Cumberland.

1760
THE TOBACCO~ROLLING ROAD

The transportation of tobacco in hogsheads rolled for a hundred miles or more o-ver primitive trails was one of the picturesque customs of our Colonial days. Be-cause dirt and water leaking through the staves of the casks could damage the tobacco it was the practice of the rollers to follow the high ground separating the watersheds and head the streams, thus avoiding wetting the tobacco at creek crossings at fords. This habit may account for many of the early meandering country roads in Virginia and other southern States.

The trails over which tobacco was hauled between the plantations and the store houses, for inspection and sale, became known as tobacco-rolling roads. There were many miles of these well-worn paths in the Southland during Colonial times and as late as 1850 when rolling was discontinued because of injury to the tobacco. The store houses were the focal points between the plantations in the interior and a shipping site on a river or other navigable body of water not more than a mile distant.

Tobacco from northern Virginia was rolled through Culpeper, Orange and Hanover counties, and Albemarle and Goochland counties, to Richmond situated at the head of navigation on the James River. The two principal rolling roads in southside Virginia were known as Cocke's Road and the Boydton Plank Road. The first joined Petersburg, where there was a water connection with the James River, with North Carolina by way of Lunenburg and Mecklenburg counties. The second ran easterly from Boydton, in Mecklenburg County to an intersection with the present location of United States Route I, south of South Hill, thence approximating closely this main thoroughfare of today to Petersburg. Another old tobacco path, which still bears the name "Rolling Road" began southwest of Baltimore, Maryland at the Patapsco River near the crossing of United States Route I.

One of the most widely publicised tobacco rolling roads, in our own times, connected northern Georgia with the nearest port on the Savannah River, downstream from the shoals a few miles south of Augusta. Although not as old as those in the Virginia Colony, this Georgia tobacco rolling road has been made famous as the theme of a book and a play. The road follows the high ridge between two creeks and never crosses any stream.

The illustration shows George Washington casually eyeing a hogshead of tobacco being rolled to his wharf at Mount Vernon, Virginia, from the plantation of a neighbor some miles in the interior. The sailing schooner in which the tobacco is to be shipped to the mother country is anchored offshore from the dock at the mouth of Dogue Creek emptying into the Potomac River. With his record in hand, George Washington has checked, as they rolled by, each one of his own casks marked, "G.W." It was not necessary to use oxen to roll his tobacco for the 700 feet between the curing barn and the flat-boat-lighter dock on Dogue Creek. The rope in the hands of the negro was used as a brake on downhill grades.

1763—THE BOSTON POST ROAD

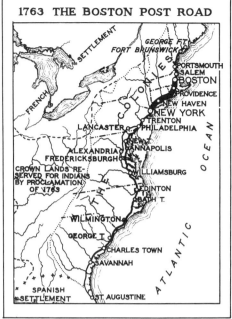

1763 THE BOSTON POST ROAD

A horseback post, by 1729, ran from Piscataway village, near Portsmouth, at the north of the Massachusetts Colony, over the Boston Post Road through New York to Philadelphia. Four weeks were required to send a letter from Boston to Williamsburg, Virginia. Improvement of this service began, in 1737, when Benjamin Franklin was appointed Deputy Postmaster General under Colonel Spottswood of Virginia, to succeed William Bradford the Philadelphia printer who had been removed from office. With Franklin at the controls the postal service soon gave promise of becoming a valuable adjunct of Colonial commerce. It was not, however, until 1753, when Colonel Spottswood died and was replaced by Franklin and Colonel William Hunter of Virginia, as associate English Postmaster General, that the service expanded to its greater extent with a line of posts from Boston to Charleston, South Carolina.

In the summer of 1753, Franklin began a personal inspection of all the post offices in the country, with the exception of Charleston. After four years of judicious reorganization he succeeded in converting the postal service into a paying institution. Franklin spent the years from 1757 to 1762 in England representing the interests of the Colonies. Following his return he made a tour of the American post offices, in the spring of 1763. He is shown in the illustration in a one-horse chaise accompanied by his daughter on horseback. A post rider is delivering an urgent message to his chief along the Boston Post Road. Franklin's daughter on the return trip rode almost the entire distance from Rhode Island to Philadelphia.

In keeping with his progressive policy, Franklin inaugurated a day-and-night post between Boston and New York, in 1764. Ten years later the aging patriot was dismissed from office by the British authorities because of his activities in behalf of American rights. Franklin was succeeded by the English Resident Deputy Postmaster General, in America, John Foxcroft, assisted by Hugh Finlay who made an inspection of the Colonial postal system for the Crown in 1773 to 1774. The Continental Congress, in 1775, rewarded Franklin with the office of Postmaster General.

The period of the War of the American Revolution measured the transition from the post rider to the mail stage coach. Prior to 1775 the bulk of long-distance travel was performed on foot or horseback and the average rate of travel was four miles an hour. Leading merchants and landowners often rode in coaches in town and even out upon the country roads but as a rule they preferred riding horses equipped with saddle bags to bumping over atrocious trails in wheeled vehicles. It was the prevailing custom for a traveler to buy a horse in a town and sell the animal for about its original cost at the end of the journey.

Before the Revolutionary War an abortive attempt was made by Jonathan and Nicholas Brown, on June 25, 1772, to operate the first public passenger-stages between Boston and New York. This service was suspended during the years of hostilities. After the conclusion of peace, Levi Pease, a New England blacksmith, restored the interrupted service, in 1784. He is credited with establishing the first successful stage line between the two cities. It passed over the upper Boston Post Road through Worcester, Springfield and Hartford.

1766~THE FLYING MACHINE

The entry of the Flying Machine stage wagon upon the run between Philadelphia and New York, in 1766, fixed another red buoy beside the main channel of highway transportation in this country. This figurative buoy marked the first feeble attempt to speed common carrier travel in America. The Flying Machine passenger vehicle was the harbinger of the present pulsing traffic along United States Route I over which roll probably more motor vehicles than on any other thoroughfare in the world. Therefore, when we compare the streamlined contours of today's automobiles with the clumsy outlines of the Flying Machine it is hard to believe that this wagon's two-day travel time between the two Colonial cities ever could have been considered rapid. The relative speed is apparent, however, when contrasted with the previous slow-moving transportation.

In 1711, when the York Road was opened throughout its length there were three main routes between Philadelphia and New York: (1) The lower, or Governor Laurie's road across New Jersey from Burlington through Crosswicks and Cranberry to Perth Amboy; (2) The old middle path, or King's highway, between the Raritan River at New Brunswick and the Delaware River at Trenton; and (3) The upper, or York Road, through Willow Grove (Red Lion Tavern), Hatboro (Crooked Billet Tavern), and Howell's Ferry (Centrebridge) to the Raritan River at New Brunswick.

There seems to have been no regular public passenger service on the King's highway out of Philadelphia, until 1725, when four-wheeled chairs were advertised to run from Three Tuns Tavern, on Chestnut between Second and Third Streets, to Frankford. One year later a petition to install a ferry below the "Falls-on-the-Delaware" was granted to James Trent whose father gave the name to Trent's town or Trenton.

The Northern Post horseback riders worked on a weekly schedule during the spring, summer and fall only, in 1750, when Joseph Borden organized the first regular stage-wagon service between the two principal cities by way of Trenton and Brunswick. The first so-called through city-to-city stage service began on November 9, 1756, along the King's highway from Philadelphia through Princetown and New Brunswick to Perth Amboy, thence to New York by stage boat.

The long ferry trip across New York Bay was displaced on April 14, 1766, by a ferry across the Hudson River at Paulus Hook installed by John Barnhill and John Masherew. They introduced the fastest stage wagon until that date, called the Flying Machine, which began a bi-weekly schedule for the two-day journey from Philadelphia through Trenton, Princetown, Elizabeth Town, Newark and the Passaic and Hackensack Rivers to Paulus Hook (Jersey City) with "the waggon seats to be set on springs" and "no water carriage, and consequently nothing to impede the journey." The comfort of passengers was served by hanging the wooden cross seats in leather straps to offset the lack of springs between the wagon body and the axles. The picture shows the passengers transferring from the sailing ferry boat at Paulus Hook and beyond the primitive skyline of New York.

1769~SAN DIEGO, CALIFORNIA

At San Diego de Alcalá, named after Saint James of Alcalá, an Andalucian Franciscan monk, the first of the Spanish missions in present California was founded on Sunday, July 16, 1769, when the Franciscan Friar Miguel José Junipero Serra blessed the Cross on a spot called by the natives, Cosoy, later the Old Town, as shown in the accompanying illustration. The building for worship and later the presidio (fort) became the most southerly of a series of twenty-one mission settlements built along a 700-mile route paralleling the Pacific Coast north as far as Sonoma, a short distance above the San Francisco of today. The second mission of San Carlos was established at Monterey on June 3, 1770. The presidio at this location was completed in 1778. The first fortified settlements, at San Diego and Monterey, were intended as refitting stations for Spanish galleons (warships) based in Manila. The government of Spain decided to occupy upper California to protect their Mexican possessions to the south following the Russian explorations in Alaska from 1745-65. Monterey, named after the then viceroy of New Spain, became the principal military, commercial and financial center of California until the region passed under United States control. All the missions along the route were visited by the Franciscans on foot. Thus this revered California road which corresponds closely with the present United States Route 101, as shown on the map, has become known as "El Camino Real of the padres."

From San Diego, El Camino Real continued southward through Arispé, Durango and Queretaro to the capital at Mexico City. This trail was the westernmost of the three main Caminos Reales which the Spaniards used to reach their buffer state to the north. The first Camino Real to be opened was that leading directly north to Santa Fé, founded 1605. From this great central road there stemmed, during the seventeenth and eighteenth centuries, eastern branches at what is now Dolores Hidalgo, Zacatecas and Durango, which joined at Saltillo, and thence ran north and east across the Rio Grande River at Presidio del Norte (Paso de Francia or French Ford), through modern Cotulla, San Antonio, Nacogdoches, Natchitoches, Natchez, New Orleans, Mobile, Pensacola, and Tallahassee to a terminus on the east coast of Florida at Saint Augustine, founded by the Spaniards in 1565, under the leadership of the Spanish naval captain Pedro Menéndez de Avilés. Any of these roads used by the Spanish Government to transmit despatches and troops were called Los Caminos Reales, meaning Royal Roads. They correspond to the Kings Highway later traveled in the British Colonies in North America. Los Caminos Reales were developed from Mexico City as a center, after 1519, when the Spanish conquistador Hernando Cortés, in command of twelve galleons and over five hundred men, landed at Vera Cruz, thence marched westward and conquered the Aztecs led by Montezuma. From that time on the Spaniards spread their dominion north and south along the mountainous backbone of the continent throughout the region now known as Middle America.

1774-THE WILDERNESS ROAD

The Wilderness Road, through the Cumberland Gap through the Allegheny Mountains, at the junction of the State boundaries of Kentucky, Tennessee and Virginia, was the main pioneer road over which poured the first waves of the great tide of migration which inundated the West. Until the War of the American Revolution, the three million inhabitants of the British Colonies in North America lived within a 150-mile belt of land paralleling the Atlantic Coast. The towering Allegheny Mountains to the west screened a vast wilderness through which roamed occasional Frenchmen and Englishmen but principally Indians and wild animals.

Prior to the Revolutionary War all land routes westward converged upon Fort Pitt (Pittsburgh, Pennsylvania), from New England and Pennsylvania, and upon Cumberland Gap from Pennsylvania, Virginia and the Carolinas. Fort Pitt was the only English outpost west of the mountains. The objective of these transmountain routes was the fertile valley of the Ohio (Indian – Oyo = beautiful) River and the land of "Caintuck" (from the Iroquois word, Ken-ta-kee, meaning "among the meadows").

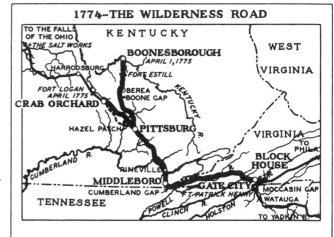

The 826-mile Wilderness Road route from Philadelphia had precedence over the 924-mile Pennsylvania route because it was the shortest all-land route to the falls (Louisville, Kentucky) of the Ohio River.

The Wilderness Road recalls the immortal name of Daniel Boone who hunted in the new country in 1767.

Boone and Michael Stoner, in 1774, were employed by Governor Dunmore of Virginia "to conduct a number of surveyors to the falls of the Ohio." They made the round trip, totaling nearly 800 miles, in 62 days. In the same year, 1774, James Harrod and his comrades, from the Monongahela-valley end of the Pennsylvania Road, floated down the Ohio and up the Kentucky rivers, and grounded their craft at the mouth of Shawnee Run where they founded Harrodsburg, Kentucky. The real beginning of the westward migration to Kentucky dates from the year 1774 when many cabins were built in the Harrodsburg and Danville areas.

The Ohio River route was the only practical path into Kentucky as late as March, 1775, when Colonel Richard Henderson met with the chiefs of the Cherokee Indians at Watauga and negotiated a treaty for the purchase of land bordering the Ohio, Kentucky and Cumberland Rivers. Daniel Boone was chosen to locate the emigrant path for the land company. With a band of well-armed men, Boone set forth from the Watauga settlement on March 10, 1775, and within three weeks had blazed a trail to Boonesborough where he began erecting a fort on April 1. The accompanying illustration shows Daniel leading a party of about forty woodsmen across Gap Creek tributary of Powell River.

Colonel Richard Henderson, with a large party, followed Boone over the Wilderness Road in the spring of 1775. One of the members, Benjamin Logan, disagreed with Colonel Henderson and separated from the party at Pittsburg thence he bore northwesterly towards Fort Harrod and established the western branch of the Wilderness Road leading to the falls of the Ohio River. This route in time became more important than Boone's path into the heart of the bluegrass region where Lexington was founded.

1794-THE WHISKEY REBELLION

1794 THE WHISKEY REBELLION

X MINGO MEETING HOUSE CRADLE OF INSURRECTION
▲ STILL ENTRY STATION TROUBLE SPOT
● WINTER QUARTERS GEN. MORGAN AND 1,500 LEFT WING MEN
▼ EXCISE INSPECTOR NEVILLE'S HOME BURNED, JULY 17
PA. FOURTH SURVEY
PITTSBURGH
BRADDOCK'S FIELD
GREENSBURG
WASHING-TON
BROWNSVILLE
MONONGAHELA RIVER
UNION TOWN
BERLIN
BEDFORD STRASBURG
FORT CUMBERLAND
ARMY LEFT WING
LEAVES FT. CUMBERLAND OCT. 25
3,300 VA. MEN UNDER GOV. HENRY LEE
2,350 MD. MEN UNDER GOV. T. S. LEE

ARMY RIGHT WING
ASSEMBLES AT CARLISLE
5,200 PA. MEN UNDER GOV. MIFFLIN
2,100 N.J. MEN UNDER GOV. HOWELL
MARCHES WEST, OCT. 10
WASHINGTON AT CARLISLE OCT. 4-11
FORT CUMBERLAND OCT. 16, LEAVES
BEDFORD OCT. 20 FOR EAST
REBELS ASSEMBLE HERE ■
MARCH TO PITTSBURGH AUG.1,
SHOW STRENGTH AND LEAVE
CARLISLE
PHILADELPHIA
(SEAT OF GOVERNMENT)
WILLIAMS PORT 20 REBEL PRISONERS
ARRIVE DEC. 25

The principal cause of the Whiskey Rebellion in western Pennsylvania, in 1794, was the lack of a good road across the Allegheny Mountains between Pittsburgh and Philadelphia. For a number of years after the peace pact of 1783, which concluded the War of the American Revolution, there was nothing but a horse path across the mountains. Salt, iron, lead, powder and other necessities had to be carried on the backs of pack horses. At the time of the insurrection, in 1794, the roads were in such an atrocious condition that wagon freight cost from five to ten dollars for each hundred pounds.

The Monongahela farmers' revolt against the Government excise tax upon the manufacture of whiskey had an economic basis. Their livelihood depended upon the sale of grain, lumber, meat, furs and ginseng. Because of the prohibitive cost of wagon transportation these products could not be hauled economically east across the mountains to the Philadelphia market nor was there any assured market if they were shipped southwest down the Ohio and Mississippi rivers. The farmers, therefore, were convinced that their only alternative was to solve their transportation problem by converting the bulky raw grain into the more concentrated form of whiskey.

It is a simple matter to convert their resentment into terms of dollars and cents. A pack horse could carry only 4 bushels of rye grain with a total weight of 224 pounds. The same horse, however, could walk along easily with two 8-gallon kegs of whiskey weighing a total of 149 pounds, with the containers, which had been distilled from 24 bushels of raw rye grain. Wagons supplied a still more profitable mode of transportation. A Conestoga wagon capable of carrying a load of one ton (2,000 pounds) could haul 36 bushels of corn grain. This same wagon could transport twenty-two 10-gallon kegs, or 220 gallons, of corn liquor. The raw grain was worth one dollar a bushel delivered in Philadelphia, and the corn whiskey, or "Monongahela rye" sold for one dollar a gallon at the same destination. The profit of converting the grain into distilled alcohol thus totaled 184 dollars for each wagon load, shown in the accompanying illustration.

The manufacture of whiskey, however, required stills which cost more than the ordinary farmer could afford. A 100-gallon still was equivalent in value to a 200-acre farm within a 10-mile radius of Pittsburgh. Because of the high initial cost, farmers in a neighborhood either clubbed together to purchase a still or else arranged to have their grain processed at a standard fee by some more prosperous neighbor who possessed the necessary capital. Distilleries thus became common in the western mountains.

Rebellion was of short duration, although it presented a threat to the stability of the newly-born Republic. President George Washington, as Commander in Chief, marched Federal troops westward over the Pennsylvania road and suppressed the movement in short order without bloodshed.

1795~THE PHILADELPHIA AND LANCASTER TURNPIKE ROAD

The Philadelphia and Lancaster Turnpike Road may be called the morning star which heralded the dawn of a new day in roadbuilding in this country. In some respects it was the outstanding highway of the thirteen original States. It was the first long-distance stretch of broken-stone and gravel surface built in this country in accordance with plans and specifications. It was the first important turnpike road in the United States, although antedated by the lesser Little River Turnpike extending west from Alexandria, Virginia. It was the most important section of the celebrated Pennsylvania Road, running west to Pittsburgh, which, following the ordinance of 1787, opened to settlement the Territory Northwest of the Ohio River. It provided cheap transportation between the populous coast regions and the "bread basket" of the newborn Republic, situated in the vicinity of Lancaster, where fertile farms yielded bumper crops of wheat. Incidentally the limestone soil, which paid such rich dividends in harvests, overlaid the parent limestone strata which, when quarried and broken into fragments, provided inexpensive local material for surfacing the road. The Lancaster Pike, and its extension, the Pennsylvania Road, was Philadelphia's bid for the fabulous wealth of raw materials, bordering the Great Lakes, in competition with the other eastern States. New York finally won with practically a water grade across the mountains following the route of the Erie Canal.

The Lancaster Pike began a new chapter in the history of roadbuilding in the United States because it was the first privately-built road of importance. It was the beginning of organized road improvement after a long period of economic confusion following the War of the American Revolution. It struck the first blow at releasing the shackles of waste and inefficiency inherent in the ancient method of building and maintaining roads, known as "statute labor," or "working out the road tax." Contrary to statements which have been given wide circulation, the Lancaster Pike was not the first macadam road in this country. As a matter of fact John Loudon MacAdam's method of road construction was not devised until about a generation later than the date when the Lancaster Pike was completed.

Construction work on the Lancaster Pike was commenced in February, 1793, and was completed practically by December, 1795.

The hospitable sign of the Spread Eagle Tavern is shown in the illustration, as it appeared in 1795, together with the stagecoach.

With the completion of the new Philadelphia and Columbia Railroad and the Pennsylvania Canal extensions to Pittsburgh, in 1834, the income of the turnpike's stagecoach and Conestoga-wagon companies suffered a drastic decline. During the next half century the road fell into disuse and lack of repair. When the automobile appeared in the Nineties the expanded economy of our prosperous country required roads free of tolls. Thus on February 25, 1902, the Court of Common Pleas of Philadelphia County dissolved the Philadelphia and Lancaster Turnpike Road Company.

1797~ ZANE'S TRACE

1797 ZANE'S TRACE

Zane's Trace was the northeastward land route beginning at the ancient buffalo crossing of the Ohio River near the mouth of Big Three Mile Creek (present Aberdeen, Ohio) and ending up the river at Wheeling connected by trails leading to Pennsylvania and Virginia. Zane's Trace, situated in the Territory Northwest of the Ohio River, was a northern segment of the long-distance path, leading to the lower Mississippi River, which later included the Maysville Pike connected by an Indian trail with the Natchez Trace. Zane's Trace was opened as an alternate cross-country route to supplement Daniel Boone's Wilderness Road over which, for more than a generation following the close of the War of the American Revolution, sturdy pioneers traveled across the Allegheny Mountains between Kentucky and the original States along the Atlantic seaboard.

Toward the close of the eighteenth century the water thoroughfare up the Ohio River from Kentucky presented such formidable difficulties that a return land trail became necessary to provide suitable transportation for the rapidly growing settlements of western Pennsylvania and Virginia. Ohio River traffic was hampered by floating ice in the winter, by spring floods, by low water in the dry summer season and by the incessant ravages of river pirates who took up their positions where the natural obstructions to navigation aided their nefarious livelihood. Only in the most favorable seasons of the year was it profitable to "pole" upstream, or "cordell" by a rope pulled from the shore, keel boats or barges laden with heavy freight. For the lighter freight and passenger traffic a land route was more convenient. Furthermore, prior to the introduction of the first steamboat on the Ohio River, in 1798, the majority of all river craft could not be propelled against the current by any reasonable expenditure of human energy. For these reasons flatboatmen were accustomed to sell their freight as well as their craft at the down-river destination and trudge homeward over a land path through the wilderness.

Following the Treaty of Greenville in 1795, the Territory Northwest of the Ohio River came into the undisputed possession of the United States Government as settlers swarmed into the new region from the eastern States. The return traffic consisting largely of produce, travelers and mail packets assumed such sizable proportions that Congress recognized the need for a 226-mile post road from the upper reaches of the Ohio River at Wheeling downstream to the most frequented western Kentucky port at Limestone (Maysville). The legislation relating to this road entitled Ebenezer Zane to locate certain lands in the territory of the United States northwest of the Ohio River.

Zane began work at once assisted by his brother Jonathan, his son-in-law John McIntire, John Green, William McCulloch, Ebenezer Ryan and others. The trail cut from Wheeling southwest to the present Zanesville followed mainly the watershed traversed by the earlier Mingo Trail; then probably retraced another Indian trail to the Ohio River. Zane's work consisted of felling small trees and widening the unsurveyed Indian Trails to accommodate horsemen and letter carriers and of constructing ferry crossings over the three principal streams.

1802—THE CATSKILL TURNPIKE

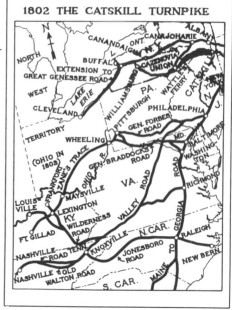

The 95-mile road westward from Catskill, New York, on the Hudson River, to Wattle's Ferry, opposite Unadilla on the west bank of the Susquehanna River, was opened to travel as the Catskill Turnpike in 1802. Within a few years after the end of the War of the American Revolution, in 1783, this route was one of the three main paths connecting New England with the Great Lakes region. According to the map of the United States, published by Abraham Bradley, Jr., about 1796, the principal New York road to the west began at the Massachusetts line at Lebanon and ran from Albany along the Mohawk River to Canajoharie and Rome to the terminus at Canandaigua. The lesser traveled western road crossed the Massachusetts line and extended through Catskill, Harpersfield, Union (Binghamton) and Painted Post to Williamsburg (Geneseo). A southwest branch of the main route connected Canajoharie with Union. Travelers from southern New York, Connecticut and Rhode Island preferred the shorter Catskill Turnpike over the older Mohawk Turnpike, opened through to Lake Erie shortly after the War of 1812-15.

Since the founding of the Plymouth Rock settlement in Massachusetts, in 1620, western migration across the present New York State had been blocked by the warlike Iroquois Confederation which resisted attempt by the New England Pilgrim Fathers to traverse their "Long House." The Indian homeland remained forbidden territory to the white man until General Wayne's victory over the red men, followed by the Treaty of Greenville, in 1795, and other treaties, opened unmolested western paths to relieve overcrowded New England. The Catskill Turnpike was one of the many roads built during the strenuous generation following the organization of the New Republic, in 1789, when internal improvements were pressed to consolidate victory by stimulating communications and commerce.

Kentucky and Tennessee had been transformed rapidly from a wilderness into settled communities with the dignity of States by 1792 and 1796 because acreage could be purchased easily from the land speculators and the later State governments. Mass migration into western New York, however, was delayed because of the unfriendly Indians and the disputed title to the region claimed by both New York and Massachusetts. After representatives of the two commonwealths reached an agreement at Hartford, Connecticut, in 1786, New York began at once to sell its holdings in large tracts in order to replenish rapidly its depleted treasury.

Wooed by the sales campaign of the land speculators, New England farmers abandoned their rock-strewn fields and moved their families to the fertile soil in the Northwest. The Catskill Turnpike and its continuation to Canandaigua were crowded with covered wagons (shown in the accompanying illustration). By 1812 there were 200,000 pioneers living in western New York. Halsey wrote about the Catskill Turnpike that, "The road ran through lands owned by the stockholders. Little regard was had for grades as travellers well knew. The main purpose was to make the land accessible and marketable.***** Ten toll gates were set up along the line,*****. Two stages were to be kept regularly on the road, the fare to be five cents a mile.***** The most prosperous period for the road was the ten years from 1820 to 1830."

1804~OLIVER EVANS' AMPHIBIOUS DIGGER

Oliver Evans' <u>Orukter</u> <u>Amphibolos</u>, which translated from the Latin means "amphibious digger," was the first steam-driven vehicle propelled on land in this country. Evans was neither the first to originate the idea nor the first to operate a steam wagon. As early as 1759 an Englishman named Dr. Robinson had suggested a steam carriage to the Scottish engineer, James Watt, who first patented his steam engine in January 1769. Watt built models of a steam wagon but did not take the idea seriously. Meanwhile, in France, Nicolas Joseph Cugnot had been experimenting with a self-propelled steam "land carriage" which he assembled and ran with some degree of success in 1769. Two years later, in 1771, Cugnot, with the financial assistance of the French Government, constructed a larger machine which upset after running about fifteen minutes upon the streets of Paris.

Oliver Evans (1755 – 1819) stated in The Weekly Register of March, 1813, published by H.Niles in Baltimore, Maryland, "About the year 1772, being then an apprentice to a wheel-wright, or waggon-maker, I labored to discover some means of propelling land carriages, without animal power. All the modes that have since been tried (so far as I have heard of them) such as wind, treadles with ratchet wheels, crank tooth, etc. to be wrought by men, presented themselves to my mind, but were considered as too futile, to deserve an experiment; and I concluded that such motion was impossible for want of a suitable original power."

"In the year 1804, I constructed at my works, situate a mile a half from the water, by order of the board of health of the city of Philadelphia, a machine for cleansing decks. It consisted of a large flatt, or scow, with a steam engine of the power of five horses on board, to work machinery to raise the mud into flatts. This was a fine opportunity to show the public that my engine could propell both land and water carriages, and I resolved to do it. When the work was finished, I put wheels under it; and though it was equal in weight to <u>two</u> <u>hundred</u> <u>barrels</u> <u>of</u> <u>flour</u> and the wheels fixed with wooden axle-trees, for this temporary purpose in a very rough manner, and with great friction, of course, yet with this small engine I transported my great burthen to the <u>Schuylkill</u> with ease; and, when it was launched in the water, I fixed a paddle wheel at the stern, and drove it down the <u>Schuylkill</u> to the <u>Delaware</u>, and up the <u>Delaware</u> to the city, leaving all the vessels going up behind me, at least, half way; the wind being a-head."

1806–LEWIS AND CLARK
AT FORT CLATSOP

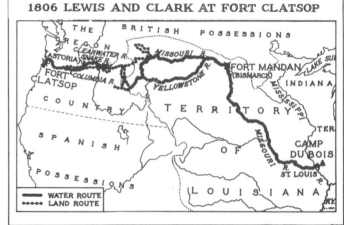

Lewis and Clark's expedition to the Northwest Coast – the first transcontinental exploration sponsored by the United States Government – followed the circuitous river routes later superseded by the more direct overland Oregon Trail. President Thomas Jefferson commissioned them to bring back information concerning the Indians and the vast territory acquired by the Louisiana Purchase of 1803, lying between the Mississippi River and the Rocky Mountains beyond which was the Oregon Country with its wealth of fur-bearing animals. As early as 1783, Thomas Jefferson, while United States Minister to France, with headquarters in Paris, had been informed of the fabulous profits in the Northwest fur trade by the Connecticut Yankee John Ledyard. This Dartmouth graduate had served as a marine corporal under Captain Cook during his voyage to the Northwest Coast begun in 1776. Jefferson arranged with the Empress of Russia for Ledyard's passage across Siberia. Thence he was to cross over into Alaska and explore the Northwest country from west to east. His expedition failed because the Empress changed her mind.

Thomas Jefferson, however, never lost sight of the immense national advantages to be gained by extending our western boundary to the Pacific Ocean. Thus after his election to the presidency he resurrected his favorite plan of an overland expedition to the West Coast. In a secret message to Congress, dated January 18, 1803, he requested authority to promote trade with the Indian tribes of the Missouri River region and to explore as far as the western ocean.

President Jefferson chose his young private secretary, Meriwether Lewis, age 28, to lead the military expedition. Lewis then selected his boyhood Virginia companion and experienced frontiersman William Clark, age 32, to be associate leader of the enterprise. Early in July, 1803, Captain Lewis left Washington for Pittsburgh, thence down the Ohio and Mississippi Rivers by boat, accompanied by 14 soldiers and 9 Kentucky hunters enlisted as army privates. En route opposite Kentucky the party was joined by Captain Clark. Because the Louisiana Territory was not transferred formally from France to the United States until December 20, 1803, Lewis established 1803–1804 winter quarters on the east side of the Mississippi River opposite Saint Louis. On May 4, 1804, the explorers set out up the Missouri River in three boats. They reached a position north of Bismarck, North Dakota, 1,600 miles from their starting point, late in October. There 1804–1805 winter quarters were erected and called Fort Mandan. The Shoshone Indian Girl, the famous "bird woman" Sacajawea and her French trader husband Touissant Charbonneau were enlisted as interpreters and guides.

Favored by spring weather, on April 7, 1805, the explorers resumed the boat journey up the Missouri River. Beyond the headwaters horses were obtained through Sacajawea's Shoshone Indian brother for the 300-mile land crossing of the Continental Divide to the headwaters of the Clearwater River down which they floated and finally built 1805-1806 winter quarters at Fort Clatsop on the south side of the Young's Bay estuary to the great Columbia River opposite present Astoria, Oregon. The return journey homeward was begun on March 23, 1806, after giving a list of the party names to the Chinook Indian Delashelwilt and hauling down the American flag, as shown in the illustration.

1808-GALLATIN'S ROAD AND CANAL REPORT

1808 GALLATIN'S ROAD AND CANAL REPORT

Albert Gallatin, Secretary of the Treasury Department, on April 4, 1808, presented a report "respecting roads and canals," at the request of the United States Senate, which became the mold from which was cast our subsequent national transportation policies. Secretary Gallatin urged, "early and efficient aid of the Federal Government" in order to "shorten distances, facilitate commercial and personal intercourse, and unite, by a still more intimate community of interests, the most remote quarters of the United States. No other single operation, within the power of Government, can more effectually tend to strengthen and perpetuate that union which secures external independence, domestic peace, and internal liberty."

The substance of the report reflects three basic concepts. First, the legitimacy of Government aid to finance transportation projects transcending local needs. In support of this premise the Treasury head advised, "Notwithstanding the great increase of capital during the last fifteen years, [The period of reconstruction and consolidation following the Treaty of Paris, in 1783, concluding the War of the American Revolution. – Ed.] the objects for which it is required continue to be more numerous, and its application is generally more profitable than in Europe. A small portion therefore is applied to objects which offer only the prospect of remote and moderate profit." Therefore, concluded Secretary Gallatin, the through routes of national importance could be financed only by the General Government because this central authority alone possessed, "resources amply sufficient for the completion of every practicable improvement." Second, that only those routes should be constructed which would yield reasonable returns upon the original investment. Because, wrote Secretary Gallatin, "It is sufficiently evident that, whenever the annual expense of transportation on a certain route, in its natural state, exceeds the interest on the capital employed in improving the communication, and the annual expense of transportation (exclusively of tolls) by the improved route, the difference is an annual additional income to the nation. Nor does in that case the general result vary, although the tolls [counterpart of the present gasoline tax on automobiles. – Ed.] may not have been fixed at a rate sufficient to pay to the undertakers the interest on the capital laid out. They, indeed, when that happens, lose; but the community is nevertheless benefited by the undertaking. The general gain is not confined to the difference between the expense of the transportation of those articles which had been formerly conveyed by that route, but many which were brought to market by other channels will then find a new and more advantageous direction; and those which on account of their distance or weight could not be transported in any manner whatsoever, will acquire a value, and become a clear addition to the national wealth." These economic principles enunciated by Secretary Gallatin nearly a century and a half ago are just as valid today. Third, a nationwide system of transportation was essential in the interests of national defense.

Written a generation before the introduction of steam railroads the recommendations of the report were restricted to public roads and canals. The conclusions, however, regardless of the instruments of transportation employed later, established the pattern which was expanded and modified as the national frontiers moved westward from the Allegheny Mountains to the Pacific Ocean.

1809—THE NATCHEZ TRACE

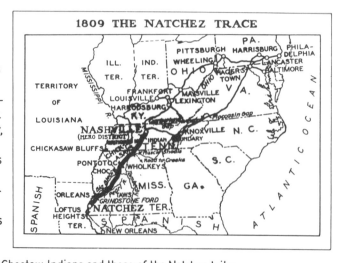

The Natchez Trace, extending from Nashville, Tennessee, to Natchez, the capital of Mississippi Territory, was the overland return route, through some 460 miles of forest wilderness and Indian lands, for the crews of the flatboats which floated down the Mississippi but could not be poled upstream. The period of its greatest usefulness began in 1798 and ended in 1817 when steamboats dominated the great river. Although the Natchez Trace was opened officially by the United States Government, in 1803, it followed a trail which had been in existence from time immemorial.

Legend has it that buffaloes, in order to evade their forest enemies, tramped this high-ridge path from the salt licks at the site of the present Nashville, Tennessee, southwest to the southern fringe of the upland where it joins the swampland bordering the Mississippi River. The Chickasaw Indians used this buffalo path as a trail from their main villages near Pontotoc, in northeastern Mississippi, to their villages near Nashville, called the Mero (Greek, meros = part or fraction) District later by our Government because of its isolated location in the western wilderness. One of the many Chickasaw trails radiating from Pontotoc led to the villages of the Choctaw Indians and those of the Natchez tribe.

In the first year of the nineteenth century, when Spain retroceded the Louisiana territory to France, the United States owned valuable lands near the mouth of the Mississippi River, access to which was blocked by intervening Indian territory. It was to provide a right of way for the flatboatmen marching homeward over the Natchez Trace, or perhaps the foreknowledge of our probable expansion toward the southwest that caused President Thomas Jefferson to direct his Secretary of War, Henry Dearborn, to negotiate with the Indians a treaty permitting the improvement of a road across their ancestral lands. Secretary Dearborn delegated Brigadier General James Wilkinson of the United States Army, Benjamin Hawkins of North Carolina, and Andrew Pickens of South Carolina, as the three commissioners vested with power to treat with the Mingos (kings), principal men, and warriors of the Chickasaw, Choctaw and Cherokee nations. By a treaty with the Chickasaws, concluded at Chickasaw Bluffs (Memphis), on October 24, 1801, the Indians, in consideration of miscellaneous goods, invoiced at $702.21, gave permission to the Government of the United States, "to lay out, open, and make, a convenient wagon road through their land, between the settlements of Mero District (Nashville), in the State of Tennessee, and those of Natchez, in the Mississippi Territory." A similar treaty was made with the Choctaw Indians at Loftus Heights (Fort Adams), on December 18, 1801, in consideration of "the value of two thousand dollars in goods and merchandise" and "three sets of blacksmith's tools." At the conclusion of these treaties eight companies of infantry troops, under the immediate command of Colonel Butler, working south from the northern Indian boundary, were ordered to cut a road to meet six companies commanded by Colonel Gaither opening the path north from Natchez.

In the illustration may be seen Captain Meriwether Lewis, Governor of Mississippi Territory, riding toward the cabin of Robert Griner (or Grinder).

1810~THE "TEAM~BOAT" FERRY

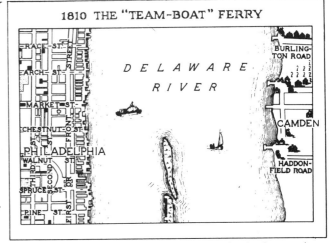

1810 THE "TEAM-BOAT" FERRY

Ferries were much in evidence during the Colonial period and the early days of the Republic for crossing wide streams too deep for fording where the cost of building a bridge was prohibitive. Even to this day ferries are resorted to where the traffic does not justify the erection of the more convenient bridge. The privilege to operate a ferry was granted by the local government under the law of eminent domain authorizing the appropriation of private property for public uses. The right to operate a ferry was obtained from the State or county by a grant constituting a contract binding the public authority and the ferry proprietor. Even owners of land abutting the stream crossed by the public road were not permitted to operate a ferry without having received such a grant. Under the contract the ferry owner was allowed to collect fixed fees in compensation for his services and the use of his property. Ferry proprietors were considered to be public carriers responsible for the life and property of the persons transported.

In early Colonial times, ferries across streams in the wilderness far removed from the centers of population were operated often by Indians with their canoes as vehicles. The native proprietor lived on one bank of a crossing.

Canoes and rafts were the most primitive types of ferries. At the more traveled stream crossings ordinary rowboats, sailboats, or wherries — shallow rowboats with seats for passengers — were employed. Horses, cattle, stage coaches and carriages were ferried on flat-bottomed barges. An inclined gangplank raised during the crossing was lowered at the landing so that vehicles and livestock could be disembarked with ease. Where the current was of sufficient strength, a cable was stretched across the stream from bank to bank. Ropes, fastened to the ends of the ferry, running through a wheel attached to the cable were so arranged in length that the ferry was held at such an angle that it could be propelled over the stream by the power of the current. At other crossings where the current was weak the ferry boat was pulled across by a rope.

The crossing of the Delaware River at Philadelphia supplies a typical example of the development of ferries at a main river crossing. "From the earliest settlement of Camden, up to about 1810, three classes of ferry-boats were in use. The smallest were the wherries, which would carry twelve or fifteen persons; and next larger were the 'horse-boats,' for the transportation of horses, carriages, cattle, etc. The principal craft were the 'team-boats,' which were propelled by horse power, [shown in the accompanying illustration.]*****The team-boats employed sometimes as high as nine and ten horses. They were arranged in a circle on a tread-wheel connected with the main shaft.

Colonel John Stevens, in 1813, built a horse treadmill ferryboat at Hoboken, New Jersey, to cross from that metropolis to New York City in competition with the steamboat monopoly granted by the State of New York to Robert Fulton. These 90-foot-long steamboats enjoyed supremacy until 1824 when Fulton's monopoly was declared unconstitutional.

1814-GROWTH OF COASTWISE TRAVEL

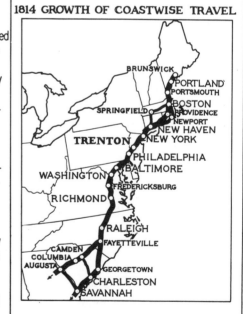

The Trenton bridge in its day was rated an engineering masterpiece. It was the second largest covered bridge built in the United States. The so-called Permanent Bridge across the Schuylkill River at High (Market) Street, Philadelphia, was the first. Before the Trenton bridge was completed, cattle on the way to the New York market crossed at the ferry a short distance downstream. The leader was loaded upon the ferry and the herd or flock swam the river behind the boat.

This crossing of the Delaware River is a reliable yardstick for measuring changes in coastwise highway travel, because of its location. The New York to Philadelphia road – now United States Route 1 – has always been, as it is today, the most heavily traveled road in this country. The Delaware River covered bridge at Trenton was the funnel through which the traffic between the two major cities had to pass. Trenton, situated at the "fall line" of the Delaware River, was at the head of navigation where the river became narrow enough for an easy crossing. Along this intersection of the piedmont plateau with the coastal plain Indian trails existed from the earliest times. One of the most important crossed the New Jersey area between the North (Hudson) River and the South (Delaware) River where a canoe ferry was installed in 1624.

During the following century river crossings so multiplied that it became necessary for the local New Jersey government to assume jurisdiction over the ferry privileges in order to assure satisfactory public service. In 1726, a petition to install a ferry below the "Falls-on-the Delaware" was granted to James Trent whose father gave the name to Trent's Town or Trenton.

During the period of reconstruction, following the War of the American Revolution, traffic swelled so rapidly that by March 3, 1798, the New Jersey legislature officially recognized the need for "a good and permanent bridge across the river Delaware." Letters patent granting a charter to the Delaware River Bridge Company were passed by the New Jersey legislature on August 6, 1803, and concurred in by the Pennsylvania Assembly. Construction of the covered 5-ribbed wooden-truss arch spans began in May, 1804. The structure was opened to traffic on January 30, 1806.

Only portions of the New York to Philadelphia road had been turnpiked (stone surfaced) when traffic began to rumble across the Trenton bridge. Proposed as early as 1792, the through turnpikes between the two main cities were not completed until about 1812. At that date a metalled roadway became of strategic importance when the British blockade halted coastwise shipping and compelled the resort to land travel up and down the coast.

As a consequence land travel increased tremendously. Long queues of Conestoga wagons rolled daily out of the northern cities towards the Southland. Passengers transferred their patronage from sailing packets to the stagecoaches running over turnpiked roads. With water commerce destroyed and the coastwise packets bottled up in the harbors by the British fleet, as shown in the accompanying illustration, the prices of market products and fuel, by May 1813, became inflated to double their usual value in the cities of Philadelphia and Baltimore.

1816—FIRST STATE BOARD OF PUBLIC WORKS

The General Assembly of the State of Virginia, on February 5, 1816, passed an act, "to create a fund for internal improvement." The act provided for the constitution of a corporate body to be called the "president and directors of the board of public works." This body was given the power to appoint a principal engineer or surveyor of public works and a secretary of the board, together with such other officers and assistants as were deemed necessary.

The board of public works created in the State of Virginia in 1816 is similar to the modern State highway commission in the following respects: (1) The directors represented various sections of the State; (2) the improvements were planned and carried out by competent and experienced engineers; (3) a special fund was set apart in the State treasury as reimbursement for the internal improvements; (4) payments were made through warrants issued by the State auditors; and (5) State aid was granted up to 40 per cent of the total cost of a turnpike project.

Loammi Baldwin, appointed in 1817, became the first principal State engineer. He was succeeded in office by Thomas Moore in 1818, who was followed by Captain Claudius Crozet, in 1822, formerly of Napoleon's French army.

The State of South Carolina entered upon an extensive program of public works in 1817, when the General Assembly created the office of civil and military engineer and gave John Wilson the first appointment. One year later, in 1818, $1,000,000 was appropriated to be expended over a period of four years in building a Statewide transportation system. This early work was devoted largely to the improvement of river transportation, because these waterways were the main arteries of travel leading into the interior.

In 1820, a board of public works took over the South Carolina program of internal improvement, which provided not only for a system of canals and navigable rivers, but also for a main State road leading from Charleston to Columbia, thence thru the Saluda Mountains.

Kentucky was another State to pass an early law establishing a board of internal improvement. The General Assembly passed this act on December 28, 1835. The legislation provided that, "the general care and superintendence, and control of all public improvements for interior communication in this State which shall belong in whole or in part to the commonwealth shall, to the amount of such interest, be vested in the board of internal improvement." The act authorized the employment of a State engineer and assistants, and the appointment by the Governor of three members, each to represent a great section of the State. It also created a separate internal-improvement fund in the State treasury.

None of these early public works organizations were restricted to roadbuilding alone. Their activities included river and canal improvement as well. A half a century of highway neglect followed the period of intense activity of these primitive State engineering organizations. The steam railroad appeared upon the scene and soon submerged all competitors in the field of transportation. The revival of highway development brought about the renewal of State-aid for roads in New Jersey in 1891, and the creation of the first modern State highway organization in Massachusetts in 1893. State aid for roads was practiced in New Hampshire as early as 1800.

1819~ THE "HOBBY-HORSE" BICYCLE

The conception of a two-wheeled vehicle propelled by the hands or feet of the operator is not new. The idea has persisted throughout the ages and vehicles of this type may be traced back to classical times and even to many centuries beyond. Winged figures astride a stick joining two wheels are illustrated on frescoes unearthed from the ruins of Pompeii. Furthermore, vehicles propelled by the muscular efforts of the occupants have been discovered also on the bas-reliefs of Egypt and Babylon. The idea recurred with increasing frequency in Europe during the seventeenth century when there was urgent need for improved facilities of transportation to satisfy the growing requirements of commerce and communication.

The origin of the "hobby-horse" bicycle has been traced by some authorities to the faded illustration of a cherub, mounted astride a vehicle, likened to a child's "Kiddie Kar," preserved in the stained glass window of the St. Giles parish church at Stoke Poges, Buckinghamshire, England. Believed to have been installed in the church in 1643, the window glass was manufactured in Italy about the year 1580.

John Evelyn, in 1665, wrote in his diary about "a wheele to run races in." The Frenchman de Sivrac contrived his célérifère, in 1690, consisting of two wheels in line rigidly connected by a wooden bar upon which sat the rider who propelled the device by pushing against the ground with his feet. Nearly a century elapsed before Messieurs Blanchard and Magurier presented a description of their vélocipède in the July 27, 1779, edition of the Journal of Paris. The vélocipède was patterned after and operated like de Sivrac's célérifère and was regarded with widespread favor. In 1784, Ignatz Trexler, of Gratz, Austria, produced a pedomotor which he claimed had the speed of a galloping horse. Perhaps he should have added downhill. M. Niepiece, a French photographic pioneer, introduced in Paris, in 1816, a vélocipède which he named a céléripède. The innovation gained considerable recognition. The good news about the vélocipède spread to Germany where Baron Karl von Drais, chief forester to the Grand Duke of Baden, made an improvement which has classed the machine as the grandparent of the modern bicycle. He pivoted the front wheel upon the frame so that the driver could steer and balance himself with the aid of the handlebar. The Baron's patent, dated 1816, claimed that the vehicle would go uphill as fast as a man could walk, could be pushed at the rate of 6 to 9 miles an hour on the level, and would roll downhill as fast as a horse could gallop. A British version of the "draisine" was patented in London, in 1818, by Dennis Johnson as a "pedestrian curricle."

The first patent for this velocipede (swift foot) in the United States was issued to W. K. Clarkson, of New York City, on June 26, 1819. The velocipede was fabricated perhaps first in this country, in 1821, by David Ball and Jason Burrill, of Hoosick Falls, New York. The American velocipede, or "hobby-horse," shown in the accompanying illustration, went out of style within a few years because of its high cost and clumsy design.

1820~DOCTOR AND CIRCUIT RIDER

On April 24, 1820, Congress passed the public-land act which permitted the purchase of tracts of 80 acres or more at a minimum price of $1.25 an acre. This measure provided a source of relief for the many families in financial straits following our first national depression brought about by speculation and the unrestrained issue of State-bank currency coupled with the deflation caused by the Bank of the United States in its efforts to regulate the excess paper money in circulation. At this time the fourth United States census showed a total population of only 9,638,453 persons, or an average of 5½ people to each square mile of territory extending as far west as the Rocky Mountains. Actually, the frontier had barely passed the Mississippi River. Daniel Boone was in the forefront of the western settlers when he established his salt works, before 1804, at Boone's Lick, now on United States Route 40 some 150 miles upstream on the Missouri River from the original French settlement of St. Louis.

Of the total population only 693,255 persons lived in villages, towns and cities aggregating 2,500 or more souls. Nearly 72 per cent of the population was engaged in agricultural pursuits which did not yield rich returns during a period when the average value per acre of farm land together with the houses, barns and other buildings was less, probably, than ten dollars, and the money in circulation less than seven dollars per capita.

There were scattered here and there in the cities and on prosperous plantations a few who enjoyed the luxuries of life and lived in affluence and ease. For each family in comfortable circumstances, however, there were many more struggling to eke out a bare existence, often in the backwoods far removed from civilization, as shown in the accompanying illustration. When tragedy overtook a pioneer family, they could rely immediately only upon their own resources. Depending upon the weather, many hours and often days transpired before help could be summoned.

The country doctor was an overworked individual who drove himself into an early grave trying to spread his limited services over too great a constituency. There were no anæsthetics. Surgery was of necessity a brutal operation and childbirth often a dreadful ordeal. Wakened in the middle of the night by the urgent cry of a neighbor who had galloped to his door, the physician would harness his horse and chaise and struggle over muddy or snow-covered trails in winter or dusty tracks in summer to bring aid to the stricken victim.

On these errands of mercy, aimed at relieving physical suffering and saving lives, the physician might be accompanied by a circuit rider, such as Francis Asbury, "the prophet of the long road," a man of God intent upon bringing spiritual solace to families in distress. Delayed by the lack of good roads their progress was interrupted more effectively by stream crossings.

1820~GENERAL JACKSON'S MILITARY ROAD

1820 GENERAL JACKSON'S MILITARY ROAD

Following the War of 1812-15, intensive efforts were made to improve our roads and canals in order to nurture the growth of the infant manufacturing industries born when the blockades of our seaports cut off imports from the English mother country. General Andrew Jackson's 516-mile military road from Nashville, Tennessee, to New Orleans, Louisiana, was projected in order to shorten by 220 miles the 736-mile distance between these two termini by way of the Natchez Trace. The right of way was cleared of timber for a width of forty feet throughout the entire length of the road. The roadway proper was ordinarily 35 feet wide but was reduced to 21 feet in width at the causeways across swamps. "There were on an average 300 men continuously employed in the work, including sawyers, carpenters, blacksmiths, etc, who were amply furnished with oxen, traveling forges, and all tools and implements necessary to its perfect execution. Thirty-five neat and substantial bridges, each measuring from 60 to 200 feet were erected, and 20,000 feet of causeway were laid. On a calculation of the pay, provisions and clothing of the soldiery thus engaged, and making a moderate allowance for the deterioration and loss of public property, we find that the general government disbursed on the occasion at least three hundred thousand dollars."

Congressional authorization, approved by President James Madison on April 27, 1816, established the terminal points at Columbia, on the Duck River, south of Nashville, Tennessee, by way of the Choctaw Agency, and Madisonville, near Lake Pontchartrain opposite New Orleans, Louisiana. The route was projected across the lands of the Choctaw Indian Nation between Columbus and Jackson, Mississippi. Uncertain as to the practicability of building the road with part of the $10,000 appropriation placed at his disposal, Secretary of War William H. Crawford wrote to Major General Andrew Jackson, with headquarters at Nashville, Tennessee, that it might be necessary to employ United States troops because of the rugged nature of the terrain and of the possible need for many bridges. General Jackson detailed Captain H. Young to obtain the information necessary to answer the Secretary of War's letter.

On March 14, 1817, Captain Young reported the completion of the reconnaissance survey as the first portion of his assignment. About a month later, on April 25, he advised the location of the proposed road noting the topography and the difficulties encountered.

The Secretary of War's apprehensions with respect to the inadequacy of the original appropriation were well founded because Congress appropriated a supplementary sum of $5,000, approved by President James Monroe, on March 27, 1818. The work was prosecuted with vigor. Major Perrin Willis was ordered by General Ripley, on March 8, 1819, to supervise the construction on the southern end of the road until its completion. Lieutenant James Scallan, of the First Regiment, United States Infantry, wrote from Baton Rouge, Louisiana, to General Andrew Jackson that the Military Road between Leaf River and the northern extremity in Tennessee was finished stating that "There has been expended on it 75,801 days' labor in three years' service by troops of the First and Eighth Infantry and a detachment of the corps of artillery, to wit: between its commencement on the 1st of June, 1817, and its completion towards the close of May, 1820."

1822—THE SANTA FÉ TRAIL

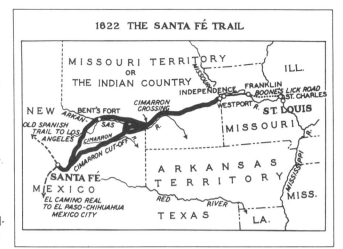

1822 THE SANTA FÉ TRAIL

The Santa Fé Trail was the first of the pioneer roads over which intercourse was established between the frontier of the United States, near the Mississippi River, and the Far West. Occasional explorers and traders passed along this historic route from earliest times but it was not until Mexico declared her independence from Spain, in 1821, that periodical and legal international trade began to flow over the Santa Fé Trail.

The Santa Fé Trail was the principal overland road by means of which friendship was cemented between the English-speaking people east of the Mississippi River and the races in the glamorous Spanish Southwest which spoke the soft, liquid syllables of their mother tongue. The trail connected at Santa Fé with the El Camino Real leading southerly through El Paso del Norte and Chihuahua to Mexico City, the hub of the Spanish empire.

Returning travelers over the Santa Fé Trail reached the East over Boone's Lick road. In 1804, when new settlers thronged into Kentucky, Daniel Boone, finding conditions too crowded for comfort, moved westward with his two sons, Nathan and Daniel, to a salt springs just beyond the Mississippi River. There he established a primitive industry evaporating the spring water to obtain salt which was floated down the Mississippi River and bartered to the French inhabitants of the village called St. Louis. Within a few years other settlers, much to the displeasure of the frontier-loving Boone, built homes near the salt springs. From this rude beginning Boone's Lick became a small black dot on the map of Missouri with the more dignified name of Franklin.

Since Boone's Lick was at the spearhead of the westward advance of the American emigrants, a wilderness path was begun, as early as 1815, from the town of Franklin to the village of Saint Charles on the Missouri River a short distance northwest of Saint Louis. The path was widened gradually until it became known as Boone's Lick road for its 150-mile length between the old salt works and the Mississippi River. This trail was used to the exclusion of all others by the pioneers.

The town of Franklin, on the north bank of the Missouri River, was really the cradle of the Santa Fé trade. By 1831, however, navigation had advanced so far upstream that a debarkation point was needed closer to the western frontier. Such a location would save more than a hundred-mile wagon haul over often muddy roads. Thus the town of Independence was founded and, by 1832, accepted generally as the outfitting station for the American caravans loading for the trip to Santa Fé. At this starting point near the Missouri River, it was customary to tighten the steel wagon tires before setting out upon the long journey across the parched plains.

In the beginning the route to Santa Fé followed around two sides of a triangle by way of Bent's Fort. The final and most direct location of the Santa Fé Trail was known as the Cimarron Cut-off. This path crossed the one hundredth meridian and extended southwest by the shortest practicable distance to Santa Fé, a total length of 770 miles.

1823~FIRST AMERICAN MACADAM ROAD

1823 FIRST AMERICAN MACADAM ROAD

The first macadam surface in this country was laid upon the "Boonsborough turnpike road" between Hagerstown and Boonsboro, Maryland. The company which built the road was incorporated by the General Assembly of the State of Maryland on January 30, 1822, and was constituted by the presidents and directors of the banks of Baltimore (except the City Bank) and of "the Hager's-town bank." The banks of Maryland as a condition for the renewal of their charters had agreed to complete this last remaining unimproved gap in the great road leading from Baltimore on Chesapeake Bay to Wheeling on the Ohio River, now United States Route 40. The work consisted of resurfacing a former county road. This section was in a sad state of deterioration, in 1821, and in winter stages required from 5 to 7 hours to cover the $10\frac{1}{4}$-mile distance. Contracts for reconstructing the road were advertised by William Lorman, the first president of the turnpike company, in September, 1822. The superintendent of construction was John W. Davis of Allegany County, Maryland. The surfacing was completed in 1823.

Mr. Davis rebuilt the old road by digging side ditches, picking and raking from the old stone roadway all large rocks and having these broken by hand so as not to exceed 6 ounces in weight or to pass a two-inch ring as prescribed by John Loudon Mc Adam of England. The first stratum of stone was "cast on with a shovel to a depth of six inches, after the manner of sowing grain." Because too much time would be required to compact under traffic the first stratum was rolled with a cast-iron roller "prepared with a box, or cart bed to carry two or three tons of stone, making weight four to six tons" and rolling until "sufficiently solid and compact to receive the second layer." After dressing off the surface "with a rake or otherwise" then "the second stratum, three to four inches thick, put on, rolled and prepared in all respects as the first stratum was, until in a state of firmness and solidity, proper to admit the third or last stratum, which can be then put on, and the surface raked and dressed off; to such shape and form as may be required, and also rolled over until satisfactorily completed." The finished surface was "15 inches deep in the centre, and 12 inches deep at each edge, and twenty feet wide," differing from Mc Adam's specification of a uniform thickness of 7 to 10 inches in his book entitled, "Remarks (or Observations) on the present system of road making," first published in Bristol, England, in 1816.

The second road surface in this country following the "Mc Adam principle" was constructed by the United States Government upon 73 miles of the National Pike, or Cumberland Road, from the west bank of the Ohio River, opposite Wheeling – at Canton (Bridgeport) – to Zanesville on the east bank of the Muskingum River. Credit for its adoption belongs to Secretary of War Barbour in the administration of President John Quincy Adams. The work was begun in 1825 and completed in 1830, under the supervision of Major General Alexander Macomb, Chief Engineer.

1825–THE ERIE CANAL

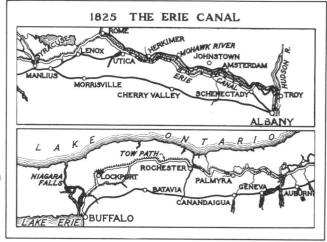

1825 THE ERIE CANAL

The opening of the Erie Canal to boat traffic on October 16, 1825, made certain the victory of New York City in the race to become the leading metropolis of America. The "Seneca Chief" was the flagship of a flotilla of canal boats which made the maiden voyage through the new waterway, from Buffalo to Albany, with Governor DeWitt Clinton, the administrative genius of the project, and other notables aboard. The fleet was greeted at the various towns along the route by colorful ceremonies, uproarious festivities, and frequent salutes of cannons and firecrackers. At the conclusion of the trip, symbolizing the junction of the western and eastern waters, Governor Clinton poured a jug of Lake Erie water into the Hudson River.

This 363-mile-long canal, with practically a water grade across the Appalachian mountain barrier, opened the shortest thoroughfare for settlement and trade between the manufacturing centers along the Atlantic Coast and the vast supply of raw materials tributary to the Great Lakes. The difference in elevation between the Hudson River, at Albany, and Lake Erie, at Buffalo, was only 566 feet and the highest intermediate elevation 420 feet above sea level. The distance to the Atlantic Ocean from Buffalo down the St. Lawrence past Montreal was 800 miles, but by way of the Erie Canal and the Hudson River, only 470 miles. The distance from Buffalo to the Gulf of Mexico through the Great Lakes and down the Mississippi River was 2,350 miles. It is not surprising, therefore, that New York City outstripped all rivals during the first half of the nineteenth century as the gateway and outlet for the West. Geography and topography destined the seven-million-dollar "Clinton's Ditch" to be the most successful engineering enterprise of its time.

The Erie Canal did not injure the business of the parallel stagecoach companies to any great extent although the canal boats bankrupted the Conestoga wagon freight carriers. There were a number of reasons for this outcome: (1) The Erie Canal was ice locked for five months of the year; (2) the stage lines were subsidized by year-round United States mail contracts; (3) the canal boats were more uncomfortable and the scenery less picturesque than some romantic authors have penned; and (4) within a few years the canal was filled to capacity with freight boats. Furthermore, the fastest packet boats averaged only a speed of less than 4 miles an hour as compared with 6 to 8 miles an hour for stagecoaches.

By 1841 there was a continuous steel rail connection from New York through Albany to Buffalo. The railroads accomplished what the canal had failed to do. They put the stagecoach companies out of business. Between 1840 and 1850 the horse-drawn passenger coaches became relics of bygone days. The Erie Canal, however, is today still a going concern.

1826-THE MICHIGAN ROAD

The Michigan Road was the main north-and-south route over which settlers swarmed into the State of Indiana. It was the overland bridge providing the shortest practicable route between the Ohio River and the Great Lakes during the pioneer period when the waterways were the principal arteries of travel. It connected Madison, on the northern bank of the Ohio River, with Michigan City on Lake Michigan. Contrary to the popular impression, the Michigan Road was not the main route over which pioneer settlers reached Michigan. The Michigan Road derived its name from the lake not the State. The immigrants to southern Michigan entered principally by way of the Erie Canal, the Great Lakes and the old Chicago Turnpike which followed the location of the earlier Sauk Indian trail leading southwest from Detroit.

The Michigan Road was used principally by the settlers who established their homes in Indiana. It was conceived by the inhabitants of the Hoosier State, built by Indiana's sons, and traversed by the immigrants bound for Indiana. Over this road the pioneers of the 1830's, called "movers," drove their ox-drawn covered wagons through the hills of the southern counties of the State of Indiana to the fertile prairies beyond the Wabash River. The track was passable during the eight months of the year when the weather was favorable but throughout the winter season it was a meandering stream of mud practically useless for travel.

The 267-mile-long Michigan Road traversed the State of Indiana by the shortest practicable route.

In the central portion of the State, the Michigan Road crossed a level plain covered with woods so dense that the rays of the summer sun penetrated rarely to the forest floor carpeted with leaf mould which retained the accumulated moisture with the avidity of a sponge. There were vast areas of swamps which the forest streams, choked with windfalls, underbrush and debris, never drained. Nevertheless, the dark, forested wilderness, the dismal swamps, the rocky and muddy stretches of the Michigan Road, the hostile Indians and the wild animals failed to deter the early settlers.

The Michigan Road was second only to the National Road as an overland route leading into Indiana. Emigrants from Pennsylvania, New England, and other eastern States floated in flatboats and barges and upon rafts down the Ohio River to begin their northward journey at the riverside settlement at Madison. Settlers journeying northward from the southern States of Kentucky, Tennessee, Virginia and the Carolinas, crossed the Ohio River from Milton to Madison.

The 100-feet-wide right of way for the road through the Potawatomi Indian lands, between the Wabash River and Lake Michigan, was obtained by a treaty consummated October 16, 1826.

In 1925, the Michigan Road handle from Michigan City to South Bend was designated United States Route 20 and southerly from South Bend to Rochester as United States Route 31. A century before, the French Canadian Pierre Frieschutz Navarre's fur-trading cabin on the east bank of the St. Joseph River, shown in the illustration, was the site for the future South Bend.

38

1827–THE NORTHWESTERN TURNPIKE

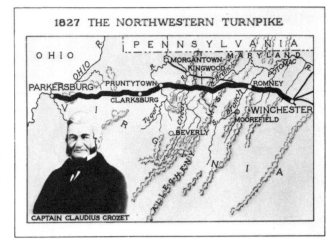

1827 THE NORTHWESTERN TURNPIKE

CAPTAIN CLAUDIUS CROZET

Incorporated, in 1827, by an act of the General Assembly, the Northwestern Turnpike was the State of Virginia's bid for the lucrative trade of the Territory Northwest of the Ohio River. Maryland's access was by way of the National Pike, opened to Wheeling on the Ohio River in 1818. The great Pennsylvania Road had been turnpiked from Philadelphia to Pittsburgh by 1820. New York outdistanced all rivals, in 1825, with the practically water-level grade across the mountains through the Erie Canal.

The need for a western road had been recognized by General Daniel Morgan and other Virginia leaders as early as 1748. In the same year George Washington probably entertained a similar vision when, as a 16-year old lad, he crossed the south branch of the Potomac River near Romney, shown on the accompanying map, as a member of a party to survey the lands of Thomas Lord Fairfax. Washington's first attempt to build a thoroughfare to the west was embodied, in 1755, in the military road cut for the ill-fated expedition of the British General Edward Braddock.

The original act of 1827, incorporating the Northwestern Turnpike, authorized subscriptions from the townspeople at Winchester, Romney, Moorefield, Beverly, Kingwood, Pruntytown, Clarksburg and Parkersburg on the Ohio River. The authors of the act made the fatal mistake of choosing a route to serve the most important towns without regard for the difficulties of the topography. Subsequently the locating engineers found a suitable grade from Winchester to Preston but encountered insuperable obstacles of terrain between that point and Kingwood. This dilemma was noised about and raised doubts in the minds of prospective stock buyers. Thereafter the project languished until 1831 when, seeking to wrench victory from the jaws of seeming defeat, the General Assembly reenacted the western road authority.

Employing every administrative precaution to avoid a repetition of the previous mistake of establishing the road through unsurveyed terrain, the Governor implemented the act by appointing an engineer of proved ability to find the best route. The chief executive made Captain Claudius Crozet, at that time State Engineer of Virginia, the chief engineer of the turnpike company. Captain Crozet had served with distinction under Napoleon Bonaparte as a French artillery officer. Later he had emigrated to this country and demonstrated his engineering talents as a professor at the United States Military Academy at West Point from 1817 to 1823. Resigning for reasons of health, he entered upon a more active outdoor life as State Engineer of Virginia, in 1824, a position which he continued to fill for nearly nine years. The route across the Allegheny Mountains chosen by Captain Crozet, shown on the accompanying map, bears the earmarks of his engineering genius to this day. The easy grades and excellent alinement developed by the survey parties under his direction more than a century ago, reproduced in the illustration, fixed the present location of United States Route 50, laid out in accordance with the most advanced engineering practices.

1830-THE MAYSVILLE TURNPIKE

The Maysville Turnpike, through the Blue Grass region of Kentucky, became a nationwide topic for discussion, in 1830, when President Andrew Jackson vetoed the bill passed by Congress to grant financial aid to its construction by the purchase of shares in the turnpike company. This veto established the policy of the National Government with respect to internal improvements of strictly local or State benefit. The veto also compelled subsequent turnpike companies and later the railroad companies to be financed as private corporations rather than as public utilities.

The issue was raised on January 25, 1827, when a resolution was passed by the Kentucky State legislature requesting congressional authorization for Federal aid in the construction of the Maysville and Lexington Turnpike Road. This thoroughfare was a 64-mile unit of the long mail road branching from the National Pike, or Cumberland Road, at Zanesville, Ohio, and running southwest across the Ohio River, through Maysville and Lexington, in Kentucky; Nashville, in Tennessee; and Florence, in Alabama; to New Orleans, in Louisiana. An appropriation bill for the construction of this turnpike initiated in the House, was defeated by one vote in the United States Senate, in the spring of 1828. Its passage would have assured the construction of the road because President John Quincy Adams favored the measure.

Undeterred, the Kentucky State legislature, on January 29, 1829, incorporated the "Maysville and Washington Turnpike Road Company" for the purpose of building an "artificial road" from Maysville to Washington, in Mason County. This 4-mile roadway, completed in November, 1830, was the first macadamized road built in the State of Kentucky. The Act authorizing this section was amended by the legislature on January 22, 1830, and the name was changed to the "Maysville, Washington, Paris and Lexington Turnpike Road Company." This Act extended the previous turnpike to Lexington and increased the capital stock.

In the Congress of the United States, a related bill passed the House of Representatives, on April 29, 1830, and the Senate, on May 15, "authorizing and directing the Secretary of the Treasury to subscribe, in the name and for the use of the United States, for fifteen hundred (1,500) shares of capital stock of the Maysville, Washington, Paris and Lexington Turnpike Road Company." "Old Hickory's" veto of the Act, on May 27, 1830, climaxed the nationwide forum.

Thus denied Federal assistance, the road was completed by town, local and State subscriptions.

Attention was focused upon the road again, in 1838, when a United States mail contractor claimed the right to travel free of tolls. Chief Justice Robertson for the Court of Appeals held that President Jackson's refusal to aid construction of the road made it mandatory for the Federal Government's agents to pay the same fees as the general public.

In the illustration the aged toll-gate keeper is portrayed in a posture of vigorous refusal as the driver of the oval-shaped stagecoach tenders what appears to be a worthless coin, a relic of the days of financial inflation following the War of the American Revolution.

1830~ THE IRON HORSE WINS

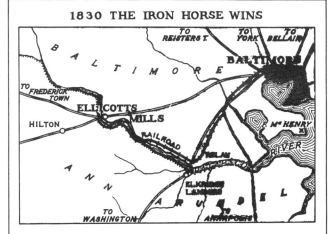

1830 THE IRON HORSE WINS

The race between Peter Cooper's diminutive Tom Thumb locomotive, on August 28, 1830, hauling a Baltimore and Ohio Railroad car filled with thirty directors and their friends, and the vehicle pulled by a gallant horse owned by Stockton and Stokes, the prominent stage-coach proprietors of that day, demonstrated for all time the superiority of steam to animal power. The steam hissing from the safety valve signalled the New Year of mechanical transportation in America.

The horsedrawn railroad coach met the locomotive, on its return trip from Ellicott's Mills, at Relay, so named as the first stage stop for changing horses eight miles west of Baltimore. Speeding side by side along the parallel double tracks, the locomotive forged steadily ahead of the horse until the leather belt broke which rotated the forced-air blower under the boiler. As the steam pressure dropped the horse galloped to victory even though the locomotive rounded the curves at a speed of 15 miles an hour and covered the 13-mile distance from Ellicott's Mills to Baltimore in 57 minutes running time.

Peter Cooper, an ingenious New York merchant, had offered to build a locomotive for the Baltimore and Ohio Railroad Company, in 1829, because he feared the loss of his investments in the lands of the Canton Company, a terminal and improvement venture dependent upon the railroad for its success. The directors had planned originally for horse operation of the railroad for the entire 384-mile distance from the Chesapeake Bay to the Ohio River. Therefore, they questioned the practicability of running a steam locomotive around the sharp curves of the track.

In view of this expert opinion, the Baltimore and Ohio Railroad directors were in a quandary. The receipts from horse-drawn power had failed to make expenses and some of the leading stockholders were considering the liquidation of their investments. It was the hope of converting losses into profits that moved the railroad management to accede to Peter Cooper's plan for introducing steam power.

Baltimore, in 1827, with a population of 80,000 people, located at the Atlantic seaport end of the great National Pike, or Cumberland Road, leading into the interior, needed the western business in order to retain its rank as the third largest city in the Union. The city's bankers had financed the construction of the road connecting at Cumberland with the National Pike which had been opened to traffic as far as Wheeling, then in Virginia, in 1818. Now Baltimoreans were disturbed by the loss of business to the Erie Canal in New York State, completed in 1825.

Goaded into action Baltimore's leading citizens laid plans for the organization of the Baltimore and Ohio Railroad, later chartered by the Maryland legislature on February 28, 1827. Only a horse railroad was considered practicable at that date because steam locomotives were still in the experimental stage even in England. Travel began on the Baltimore and Ohio Railroad on January 7, 1830, when paying passengers were carried to the far side of the Carrollton stone-arched viaduct. Later wind-driven sailing cars and horse-powered treadmill cars were tried without success. Thus began the first common-carrier railroad enterprise in this country. All horses were replaced by steam locomotives on July 31, 1831.

1836~EL CAMINO REAL

El Camino Real, the Royal Highway, comparable in importance with the King's High-way of the English, was the name given by the Spanish conquistadores to any of the routes used for internal communication, or for the defense of their far-flung dominions in North and Middle America from the border encroachment of France, England and Russia. Following the landing of Hernando Cortés, at the present Vera Cruz, in 1519, the Spaniards captured Mexico City and thence pushed north and south along the highlands of the Cordilleran mountain backbone of the Continent. Although the conquerors used a vast system of roads extending in many directions, there were three main roads .lead-ing north from their capital at Mexico City.

The first of these great Los Caminos Reales, developed during the sixteenth cen-tury, led north to the present Queretaro, Zacatacas, and Durango, through the Chihua-hua desert, crossed the Rio Grande at El Paso del Norte (Northern Ford), and followed this river to Santa Fé, New Mexico, founded about 1609, at a time when the English were establishing a foothold in Virginia. From this central El Camino Real there stemmed,

during the seventeenth and eighteenth centuries, eastern branches. A northwestern branch was opened about the time of the War of the American Revolution.

As the conquistadores advanced to conquer new territory, they left behind settlements constituted by three main buildings, namely: the mission, the center of ecclesiastical services; the presidio, or military fort; and the villa, or pueblo, the town house for the civil government. The historic building known as the Alamo, now a patriotic shrine in Texas, was a chapel of the old Spanish mission at San Antonio. The ruins of the structure are shown in the accompanying illustration as it appeared after the bombardment by the troops of the Mexican General Santa Ana, in 1836. Within the walls of the en-closure, led by Colonels William B. Travis and David Crockett, a band of some 88 gallant Texans made their "last stand" against an overwhelming force of 2,500 Mexican soldiers. The name Alamo, meaning poplar or cottonwood in Spanish, had its origin in a nearby grove of trees of that species.

At the time of the massacre the roads were indescribably bad and vehicles were crude. Covered wagons were drawn by oxen, rather than horses or mules, because these slow-moving animals could withstand greater hardship, pull heavier loads, and cost less to replace. The carreta cart, with its solid-wood tympan wheels, which were rarely greased and consequently screeched and creaked, was the favorite vehicle of the Mexicans. Pack mules were resorted to on roads impassable for wheeled vehicles. The saddle horse was a favorite means of travel.

1836~THE DESERTED VILLAGE

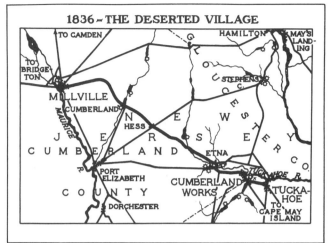

1836~THE DESERTED VILLAGE

Deserted villages in rural areas, illustrated herewith, like the later "ghost-town" remnants of once prosperous mining camps in the Far West, were the products of changing conditions in agriculture, industry, transportation and other phases of the social life of a community. The effect of the loss of raw materials upon a small mill town was described in the Scientific American magazine on October 6, 1849, under the heading of "A Deserted Village" as follows: "Nearly halfway between Millville and Tuckahoe, N.J., the traveller suddenly leaves the almost interminable waste of stunted pine and oak, the long sandy road, and the oppressive heat, and, as if by magic, a romantic hamlet, nestling beside a lake, bursts upon the view. Here he may rest his jaded horses beneath the overhanging willows and enjoy the scene to his heart's content. The village is known as 'Cumberland Works,' and consists of about twenty-five cottages, with several spacious buildings, once occupied as Mills, Iron Foundries, Forges, etc. But a deep and impressive silence now hangs over the place; the tenements are dilapidated and leaning as if ready to fall to the earth. The water-wheels are motionless; the furnaces are no longer glowing; the trip hammer, that great heart which once beat night and day, has ceased its pulsations, and all save beauty has departed. The Iron Works were formerly conducted by Edward Smith, Esq., of Camden, but owing to the rapid decrease of timber in the neighborhood, were abandoned some thirteen years since."

The depletion of raw materials characterized by this citation has been a prolific reason for the abandonment of settlements which have grown up around saw mills, grist mills, metal and coal mines and similar activities. The substitution of steam power for water-wheel-driven mills and the unreliable windmill, in the early part of the nineteenth century, however, introduced the Industrial Revolution.

"The Fate of the Rural Town" under the impact of the steam railroad, which minimized distance, was recounted in the May 11, 1895, issue of the Scientific American, to-wit: "The population of the whole country has immensely increased, [from 15,388,080 in 1836 to 69,471,144 in 1895] while scores and hundreds of the rural towns have steadily declined in population and wealth. In view of these facts, we must look for a deeper cause, and that cause we find in the new facilities for travel and transportation. The railway is an immense centralizing power. ***** In its presence all things pass and the whole world is made anew. The immediate results from the introduction of steam as a motive power were felt long ago; the remoter consequences are now being revealed in every cause and in every line of business. The change is nowhere more clearly seen than in the relation of the inland town to the commercial metropolis. When men reached the interior by horse power, by the ox team, or on foot, the rural town had a living chance to advance in population and wealth. For the industrial army which had moved into the wilderness or the open country, the rural village was the new base of supplies. The commissariat must go along with the columns. The large center was too far away. But the coming of the railway abridged distance. It brought the village ten or twenty miles away in touch with the great city, making it a sort of suburb. The outlying depot of supplies is no longer needed; the railway train has taken the place of the country storehouse."

1836~THE NEW ENGLAND TOWN HALL

Within the confines of the "town hall" or "town house," shown in the accompanying illustration, were assembled the New England "town meetings" eligible to male citizens who were twenty-one or more years of age. These citizens were residents of the area called the "town" or "township" which included not only the village but also the tributary farm, waste, pasture and other lands. The "townships" consisted mostly of odd-shaped pieces of land selected, perhaps, by groups of Puritans who had migrated to America to escape the restrictions to their freedom of worship then prevalent in their English homeland. It was logical, therefore, that these faithful pioneers should cross the ocean not simply as individuals but as church congregations under leadership of such beloved pastors as John Cotton or Thomas Hooker. Upon their arrival in New England, after making arrangements with the authorities for a grant of land, the Puritan fathers cleared an opening in the wilderness and built their log-cabin homes within comfortable walking distance of their central "meeting house" where they gathered every Sabbath day for worship. Nearby the congregation's "meeting house" was situated the "common" pasture — for general use, the school house, and the crude fort, or blockhouse. The clustered cabins promised a ready shelter in the event of a surprise Indian attack.

As the community grew to the dignity of a village, around the "common" greensward, there were added from time to time a general store, perhaps a tavern, and a "town hall" for the transaction of public business. Hitherto the "town meetings" had been held in the congregational "meeting house" until the population multiplied to a number warranting the erection expense of a separate "town hall."

The "town meeting" was convened once each year, usually during February, March or April. Because every adult male townsman was entitled to participate, the "town meeting" was the simplest and most democratic form of local government ever devised. The "town meeting" was a direct descendant of the ancient clan, an aggregation of family groups controlled by a "headman." The word town may be traced back to the primitive wooden stockade, called a "tun," which surrounded an English village.

The responsibility for public road improvement, a primary function of the township authorities since the early days of the New England settlements, began to be shared with privately-owned toll road companies shortly after the War of the American Revolution. The widespread financial failures of these turnpike companies, the emergence of the steam railroad, by 1836, as the preferred mode of transportation, and the subsequent withdrawal of the Federal Government from extensive roadbuilding, handed back the old burden of wagon road and bridge construction and maintenance to the New England townships, to the Louisiana parishes and to the counties and local authorities in the other States.

1836~THE PARISH CHURCH

1836~ THE PARISH CHURCH

The parish church, governed by its vestry and cared for by its church wardens, was the cradle of all social intercourse and government in early Virginia. Near the church, after the Sunday morning service, the women might gather in groups to talk over the current events such as weddings, social engagements and what not, while men discussed the business of the parish, the state of the London tobacco market, or perhaps Colonial politics. The parish church was the symbol of the customs of the people of Virginia just as the meeting house was a reflection of the life and habits of the Puritan fathers and mothers in New England. The principal difference in the social structure of the two regions found expression in the character of the church government. The New England meeting house was founded by Puritan separatists from the established church in England who had migrated to America in order to gain freedom to worship according to the dictates of conscience. Each congregation, therefore, was considered a distinct entity governing itself and not responsible to any superior church authority. In Virginia, on the other hand, the parish church in America was patterned after its Anglican counterpart in the old country. The Virginia Cavaliers, who were English country squires for the most part, simply transplanted to the Colony the kind of church authority with which they had been always familiar.

It was the custom in England for the parish to include only as much territory as could be served conveniently by a single priest. The English parish has been described aptly as "locus quo populus alicuius ecclesiae degit" – "the multitude of neighbors pertaining to one church." This de-description fits also the Virginia parish as well as the New England town. As a matter of fact, the words parish and town were employed interchangeably in the early records. Sometimes the parish was so large that it might coincide with the area of an entire county. Elsewhere, perhaps, a county might cover two or three parishes.

The Virginia parish vestry, so called because they originally met in the cloak room of the church, were the administrative body of the parish, responsible men selected to administer ecclesiastical and often civil duties. In the latter respect, the parish was overshadowed by the county.

In only one of the southern States – Louisiana – has the parish subdivision assumed the status and duties of the county. In that State, dating from 1807, the parishes were delimited from early divisions of the region made by the Spaniards for religious purposes. As late as 1918, the Louisiana parishes were comparable in size, organization and administrative power with the counties in other States.

The parish, the township and the county organizations became once more responsible for the main public roads when the steam railroads gained the ascendency.

The rural church, shown in the accompanying illustration, is a lineal descendant of the parish church of the South and the New England meeting house.

1836—THE COUNTY COURTHOUSE

1836 THE COUNTY COURTHOUSE

The county courthouse, shown in the accompanying illustration, situated at the county "seat," or capital, was the most important center of local government in the Southern and Western States. In Virginia, where the first counties were established in Colonial America, the court day was not only a well advertised date for conducting public business, but a holiday as well for the farmers and their families. Large numbers of people from every direction swarmed from the countryside into town, traveling in wagons, buggies, on horseback or afoot. On the steps of the courthouse, in the corridors and on the green in front were swapped stories and gossip, debts were paid and new obligations contracted, horses and property were auctioned and political speakers harangued the crowd from the "stump."

The word county is derived from the territory in France governed by a count. After the Norman conquest of England in the eleventh century, this same name was applied to the English shires, the antecedent of which were the ancient tribal governments superior to the clans, or family groups, which had their habitat in the towns. These two forms of English local government—the tribe, or county, and the clan, or town—were transplanted to the English Colonies in America.

The town predominated in New England because of the compact population groups, while the county proved more workable for the people scattered over large plantations in the Southern States. The exception—the parish—became the accepted political subdivision in Louisiana. In New England, the Puritans migrated as church congregations, settled the land in groups and it was natural that they should gather in towns for business and social intercourse. Virginians, on the other hand, lived on plantations of vast extent, the better to accumulate profitable returns from their staple crop—tobacco. The land was granted to individuals and was sometimes an enormous area. For example, "John Bolling, who died in 1757, left an estate of 40,000 acres, and this is not mentioned as an extraordinary amount of land for one man to own."

These plantations were situated often a long distance from the coast, for Virginia was penetrated by a number of navigable streams. English ships could sail upstream, deliver their merchandise at the plantation wharf and receive in exchange a cargo of tobacco. The plantation owners raised practically all the necessities of life so there was no need for the development of towns as depots of exchange between the seller and the buyer.

In Virginia, as early as 1769, the County Court, composed of eight or more gentleman inhabitants, was commissioned by the Governor of the Colony. Besides its judicial functions, the County Court was responsible for the conditions of the highways, causeys, bridges and "church roads." The County Court could enter into a contract or perform the work by day labor. The laborers were recruited by subdividing the county into precincts, or "walks," each in charge of a "surveyor" or foreman. At the times prescribed by the County Court, all "tithable" males were required to work under the direction of the "surveyor." "Tithable" persons were local residents over 16 years of age, whether free, slave or indentured, white women excepted. Owners of two or more "tithables" could send them as substitutes in lieu of working in person.

1839~OUR FIRST IRON BRIDGE

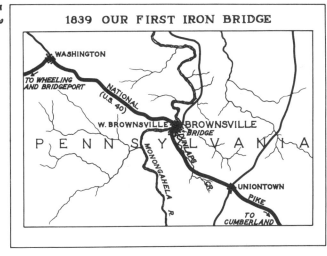

The first iron bridge in the United States was completed in 1839, over Dunlap's Creek (formerly Nemacolin's Creek) along the present Main Street in Brownsville, Pennsylvania, on the route of the National Pike, or Cumberland Road, leading to Bridgeport. There had been a succession of bridges at this location. One of these was a chain bridge of the type patented by James Finley This structure suspended partially over the stream and abutting shores, at a height of 25 to 30 feet, collapsed with a thunderous crash early in the year 1820 because of a weight of snow exceeding its bearing capacity.

The first cast-iron bridge was built as one of the several structures requiring reconstruction in the repair of 131 miles of the National Road by the Federal Government from Cumberland on the Potomac River to Wheeling on the Ohio River prior to the relinquishment of the road by the Federal Government to the States of Maryland, Pennsylvania and Virginia for future maintenance. Captain Richard Delafield, of the Corps of Engineers, United States Army, who had been placed in charge of the reconstruction of the road in 1832, conceived the idea of an iron bridge because of the proximity of the foundries at Brownsville. A number of cast iron bridges had been built in Europe before this date.

The abutment and wingwalls of the bridge were built of sandstone. The abutments are 25 feet wide across their faces, 14 feet thick and have an average height of 42 feet. The arch is composed of five ribs spaced 5.77 feet center to center, consisting of nine cast-iron hollow elliptical voussoir sections, bolted together, upon which rest the open cast-iron spandrels supporting the floor of the bridge. The arch has a span of 80 feet and a rise of 8 feet. The original Mc Adam surface, 1½ feet in thickness, was confined at the sides by road sustaining plates 1 foot 6 inches.

While the construction of the bridge was in progress Captain George Dutton, on August 8, 1838, assumed the supervision of the repair of the Cumberland Road succeeding then Major Delafield who had been appointed Superintendent of the United States Military Academy at West Point, New York. On October 15, 1839, Captain Dutton reported to Colonel Joseph G. Totten, Chief of Engineers, United States Army, with headquarters at Washington, D.C., that the cast-iron bridge over Dunlap's Creek, had been completed on July 4, of that year.

For more than a decade following its completion, a steady stream of stagecoaches and heavy Conestoga freight wagons rolled over this substantial bridge until January 10, 1853, when the Baltimore and Ohio Railroad was opened to Wheeling on the Ohio River. Thereafter, stage and freight lines steadily disappeared from the National Pike, or Cumberland Road. Then followed nearly a half a century of disuse of the road until the introduction of the automobile. Since the beginning of the twentieth century a procession of automobiles and trucks has raced across this historic cast-iron bridge at high speeds carrying heavy loads never anticipated by the designing engineers.

1840 – THE NATIONAL PIKE

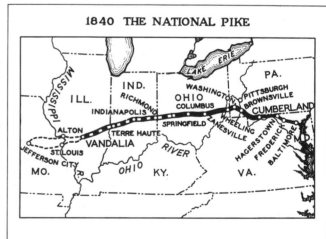

1840 THE NATIONAL PIKE

The National Pike, or Cumberland Road as it is named by statute, was the first important road to be built with Federal funds in this country. Now known as United States Route 40, the commissioners appointed by President Thomas Jefferson selected the location following the old portage path across the Allegheny Mountains from tidewater on the Potomac River at Cumberland, Maryland, to the nearest practicable head of navigation on the Monongahela branch of the Ohio River at Red Stone Old Fort (Brownsville, Pennsylvania), 72 miles away. From the day, in 1818, when travel began as far as Wheeling, then in Virginia, the National Pike shared with the Pennsylvania Road the bulk of the east-west movement until the Baltimore and Ohio Railroad reached the Ohio River in 1853. Rising and falling over the hills in the States of Maryland, Pennsylvania, Virginia, Ohio, Indiana and Illinois, along a series of straight courses of greater length than on any other road in America, this great thoroughfare spanned a distance of 677 miles between its point of beginning at Cumberland, to the projected terminus at the Mississippi River. The National Pike was the westward extension of the 135-mile turnpike from Baltimore through Frederick to Cumberland. In the heyday of its career, between 1820 and 1840, this long road was thronged with lines of heavily-laden Conestoga freight wagons passed continually by fast stage coaches drawn by six galloping horses.

Congress in 1803, one year after Ohio was admitted into the Union, agreed that a "2 per cent fund," derived from the sale of public lands, should be devoted toward the construction of roads TO and THROUGH the new commonwealth. The National Pike was the result of this "compact."

On March 29, 1806, Congress passed an Act authorizing the construction of this road, from Cumberland, Maryland, to the State of Ohio. Funds from the National Treasury were to be reimbursed from the State "2 per cent fund." This financial plan turned out to be a pious hope that was never realized. By 1808 the right of way was cleared one-half width as far as Brownsville, Pennsylvania. The first road construction contract was let on May 8, 1811, and the partially completed highway was opened to traffic as far as Wheeling on the Ohio River, in 1818.

When Indiana, in 1816, and Illinois, in 1818, were admitted to Statehood, the Federal Government entered into roadbuilding "compacts" similar to that with Ohio. Under these agreements for repayment, which never were carried out, Congress in 1820 appropriated funds for continuing the National Pike through the States of Ohio, Indiana and Illinois to the Mississippi River. By 1830, the location reached Vandalia, then the capital of Illinois, but never progressed beyond that town. Already the 18-year struggle had begun between St. Louis, Missouri, and Alton, Illinois, to win the prized crossing of the Mississippi River.

By the time that the new stone bridge was completed by the United States Army Engineers across Wills Creek, at Cumberland, Maryland, in 1836, the shadows of eastern railroad competition had lengthened toward the west. The National Pike at its eastern extremity began to feel the loss of business to the Baltimore and Ohio Railroad, in 1840. The stage coaches stopped running east of the Ohio River, in 1853, when the railroad reached Wheeling.

The last construction expenditure on the National Pike was made in Indiana, in 1841.

1840~SHENANDOAH VALLEY TURNPIKE

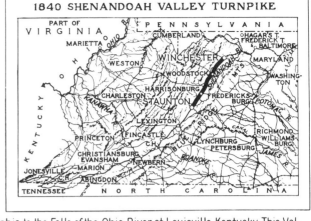

1840 SHENANDOAH VALLEY TURNPIKE

The picturesque Shenandoah Valley, some 150 miles long and 10 to 20 miles wide, is the northern extremity of the Appalachian Valley, or Valley of Virginia, which extends for 360 miles across the western portion of the State from the Potomac River south-westerly to the boundary of the State of Tennessee. This great groove between the Appalachian Mountains on the west and the Blue Ridge Mountains on the east is drained by five rivers of which the most northerly is the Shenandoah River debouching into the Potomac River.

A main Indian trail traversed this valley throughout its entire length from the earliest times. It was used as a war path by the Delaware and Catawba Indians and later, widened to accommodate the wagons of the settlers, became known as the Wagon Road. This path was the principal land road over which settlers reached Daniel Boone's Wilderness Road leading through the Cumberland Gap to the promised land beyond the Appalachian Mountains. According to Filson, the distance was 826 miles from Philadelphia to the Falls of the Ohio River at Louisville, Kentucky. This Valley Road was preferred by settlers to the uncertain and perilous water route down the Ohio River.

The northern extremity of the Valley Road became known as the Shenandoah Valley Turnpike or Valley Turnpike after 1840 when the company completed the macadam surface. On March 3, 1834, the General Assembly of Virginia passed an Act incorporating the Valley Turnpike Company.

The Act covered a distance of 92 miles between Winchester in the north and Staunton to the south. In response to the request of Bushrod Taylor, the first president of the company, the Board of Public Works assigned I. R. Anderson, as principal engineer of the project.

During the War between the States the Valley Turnpike was fought over many times by the Federal and Confederate armies. As a result the road suffered serious damage and the company collected negligible revenue because "of the Army destroying bridges, injuring Toll Houses, and we are getting very little tolls."

Conforming to the customary pattern for toll road companies the Valley Turnpike was in financial difficulties during the major portion of its existence. In 1841 the company borrowed $25,000 from the State. Toll collections never were sufficient to pay the operating and maintenance costs of the property not to speak of retiring this debt while yielding a reasonable return on the original investment. The Valley Turnpike expired officially on March 20, 1918, when it was transferred to the State highway system by an Act of the State Legislature.

The Valley Turnpike supplied the locale for Thomas Buchanan Read's well known poem describing Union General Philip Sheridan's ride in pursuit of Confederate General Jubal Early, on October 19, 1864:

> "But there is a road from Winchester town,   A good, broad highway leading down:
> * * * * * * * * * * * * * * *   With Sheridan fifteen miles away."

1843 – THE OREGON TRAIL

1843 THE OREGON TRAIL

The Oregon Trail was the first overland wagon road to the Pacific Coast along which settlers trudged, rode, and sang in the greatest tide of travel that ebbed and flowed in pioneer America. Beginning at Independence, Missouri, the Oregon Trail climbed steadily for 921 miles to the South Pass over the Continental Divide, and reached its highest elevation (8,200 feet) a short distance beyond Fort Bridger. Thence the road ran downhill to Oregon City, 2,000 miles from Independence, except for the 5,000-feet-high Blue Mountains beyond Fort Boise and the 4,000-feet-summit of the Barlow Road south of Mount Hood.

The Oregon Trail was conceived in the womb of time during the period from 1500 to 1792 when sea explorations were made by the Spaniards, French, Portuguese, Dutch, Russians and Americans to discover the Northwest Coast of America. In 1792, Captain Robert Gray of Boston, in his good ship Columbia Rediviva, discovered and named the great river.

Land explorers, seeking fur-trading profits, searched for the best route to the Northwest Coast from 1745 to 1833.

During the ensuing decade, from 1834 to 1843, fur traders yielded control of the Oregon country to missionary colonists and farmer emigrants. In 1834, Reverend Jason Lee led a party over the faintly discernible trail to Fort Vancouver, the new headquarters of Hudson's Bay Company. Dr. Marcus Whitman, a Presbyterian missionary, drove the first wagon to the Columbia River in 1836. Dr. Elijah White led the first large group of emigrants over the Oregon Trail in 1842, when the Webster-Ashburton Treaty between the United States and England and hard times in the East created conditions favorable to the Great Migration of 1843 which marked the real beginning of the Oregon Trail.

Thousands of emigrants, swarming to Oregon over the now well-beaten track and down the Columbia River on rafts, as shown in the accompanying illustration, clinched the American claim to the Northwest country from 1844 to 1848 when the cry of Gold! deflected the stream of travel near Salt Lake toward the new bonanza in California. In 1846, the devotees of peace hushed the raucous warcry of "Fifty-four-forty or fight" as the Oregon Question was settled by England's relinquishment of all Oregon land claims between latitudes 42 and 49 degrees. The California gold seekers as well as those bound for the Oregon gold fields crowded the trail from 1849 to 1852. Meanwhile, in 1850, the Donation Land Act attracted homeseekers once more to Oregon. The next important series of events included the Statehood of Oregon in 1856, the consolidation of the overland stage lines in 1858, the pony express riders galloping over portions of the Oregon Trail in 1860-61, and the eclipse of the old trail by steam railroad to the Pacific Coast in 1869. Since then, motor vehicles have transformed the growing stream of travel and United States Route 30 has become the new name for much of the Old Oregon Trail.

1844—RED R.... X CARTS

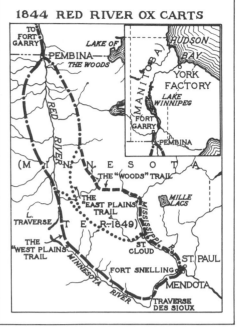

1844 RED RIVER OX CARTS

Red River ox carts were the .... ion in the Territory and State of Minnesota during the quarter cer..... ...... carts were made without i-ron or nails, the tires even being r..... ...de with shafts and one ox to a cart, with harness much like ho..... ....tly of rawhide. The axles are never oiled, and in driving each ....which can be heard on a still day or night for miles. There ....ese were the freight cars carrying goods west." So wrote A.W.Gr..... ....wer Red River settle-ments in 1869.

The main routes followed by the R..... .....ver Valley with the head of navigation on the Mississippi Rive..... .....nnesota or St.Peter's River, Fort Snelling was erected by the G..... .....a, on the east bank of the Mississippi, was at that time the entre..... .....inhabitants of the Red River Valley–an almost level plain some..... .....g, from the head-waters at Lake Traverse northerly into the ou..... .....on Bay in Can-ada. The fertile Red River Valley was the hab..... .....mals including the wild buffalo, beaver, otter, mink, fisher, mart..... .....r, the prolific buffalo rapidly disappeared. From 1763 to 1815, ...ers, the pelts from the Red River Valley were transported nort..... ..... of the Hud-son's Bay Company thence by the long circuitous s..... .....d North At-lantic Ocean to England.

The direction of the Red River fur trade began ....e, about 1820, as the growing number of settlers in the Red River Valley, between Pembina on the Canadian bord..... .....ssiniboine River, found it more convenient and profitable to deal with the business concerns of the United State..... .....x carts began to travel the trails in large numbers be-tween the Red River Valley and St.Paul which then n..... .....ding center.

By the time that Minnesota Territory was organize..... .....hed between the headwaters of the Red and Mississippi Rivers. The three principal routes were designated as th..... .....and the "west plains" trail, as shown on the map.

With their regular schedules interrupted by the Sioux..... .....r ox carts gradually faded from the transportation scene following the introduction of steamboats on the Red..... .....rior transportation supplied first by the Burbank and Company wagon freight lines and later by the steam r.....

1846~THE PLANK ROAD CRAZE

The first plank road in the United States was opened to traffic on July 18,1846, from the city of Syracuse to the foot of Oneida Lake, in the State of New York. The road was built by a privately-owned toll corporation known as "the Salina and Central Square Plank Road Company." The advocates of this new type of surface made extravagant claims as to its superiority over macadam. They held that it would be relatively inexpensive and easy to maintain and that for smooth-riding qualities it had no peer. In spite of the reluctance of the engineering profession to endorse any material that by its nature was transient in character, this new road surfacing struck the popular fancy and during the following decade thousands of miles were built in many States until the disillusioned public began to appreciate the fact that the life of any road is limited by the lasting qualities of the material of which it is built. It took about 10 years to demonstrate this axiom – just long enough for the wooden planks to rot away and wear out. The plank-road type was introduced into this country from Canada where about 500 miles were laid between 1834 and 1850. It was especially adapted to that area where standing timber was abundant and the work of making planks provided work for small local saw mills.

The Salina and Central Square Plank Road, was built originally with a single track eight feet wide, shown in cross section above. The first track was laid level transversely until it was discovered that rain water percolating through the cracks, puddled in the trough beneath the planks, softened the sub-grade soil or floated the planks because the entrapped water had no way of escape. It became necessary, therefore, to cut drains through the earth shoulders connecting with the side ditches. When the second parallel track was placed the planks were inclined three inches transversely so as to carry the surface water directly into the side ditches. The planks were laid also in groups with staggered ends in order that the wheels of a wagon which had left the road to pass another vehicle would be provided with a projection upon which to remount the planking instead of scraping along a smooth edge and cutting a deep rut alongside. Although the cross-planking was four inches in thickness and made of hemlock, a great number were broken under the heavy loads of cord wood, timber and other material shown in the illustration. The principal reason for this breakage was that the original foundation lengthwise stringers consisted of light timbers spaced too far apart with the result that the center of the track gave way under the excessive loads. This defect was corrected when the second track was built by using stringers made of two pieces of hemlock, each three inches deep by six inches wide, with the joints staggered so as to be opposite the middle of the parallel stringer. Since the "track", or distance between the wagon-wheel centers in New York State averaged 4 feet 8 inches, the stringers were located three feet apart to provide the support for the loads concentrated upon the wheels. The earth between was packed hard to give extra stability to the cross planking. The planks were nailed crosswise at right angles to the center line of the roadway because it was found that traffic tipped and loosened planking fastened upon the diagonal.

(blackboard diagram labels:) W W W=weight $\frac{2}{3}h$ h a b c a' Horizontal thrust Wrought iron diagonal Cast iron vertical members = thrust stress

$$\text{al pressure} = 2W$$
$$2W\frac{ab}{\frac{2}{3}h} = \frac{2W}{\frac{2}{3}h} = \frac{3W}{h}$$

1847–RATIONAL TRUSS–BRIDGE DESIGN

The rational design of bridge trusses dates from 1847 when correct methods were devised for calculating the stresses and strains induced in truss members by known deadweights and moving loads, coupled with experimental data with respect to the strength of the bridge materials. In that year a civil engineer named Squire Whipple, who made mathematical and philosophical instruments in Utica, New York, and at one time was employed in designing bridges for the Baltimore and Ohio Railroad, published "A Work on Bridge Building Consisting of Two Essays," relating to wooden and iron bridges. Concerning this book, Whipple later affirmed, "It is believed that no previous attempt had been successfully made to reduce Truss-Bridge construction to its simplest elements, and to determine by exact calculations, the forces acting upon the various parts of such structures and to deduce thence the proper sizes and proportions of such parts upon known and reliable principles."

1847–RATIONAL TRUSS–BRIDGE DESIGN

SQUIRE WHIPPLE   HERMAN HAUPT

By one of those unexplained coincidences believed by some to be induced by a climate of parallel scientific thought, Herman Haupt published in 1851, from a manuscript completed in 1846, a book entitled, "General Theory of Bridge Construction" in which he also explained methods for "calculating the strains upon the chords, ties, braces, counter-braces, and other parts of a bridge or frame of any description." Haupt began the studies which culminated in this book, in 1840, when he "commenced a course of experiments on models" while engaged as a civil engineer in the construction of the Pennsylvania Railroad.

Neither Whipple nor Haupt originated the basic principle of truss-bridge design which recognizes the triangle as the only rigid geometrical figure – unable to be distorted without deformation of a component part. This fundamental fact was indicated in the notes of the Italian Leonardo da Vinci as early as 1500. Evidence of this knowledge is contained in the works of the Italian bridge builder Andrea Palladio (1518-1580).

Nor did Whipple or Haupt originate the processes of mathematics or mechanics by which they analyzed bridge trusses. The calculation of the forces applied by lever arms of varying lengths had been accomplished by the celebrated Greek geometer and mechanic, Archimedes (circa 287–212 B.C.). During the sixteenth and succeeding centuries the resolution of forces by the parallelogram, the theory of beam and column flexure, was known to Galileo, Michelangelo, and Euler. John Millington in his "Elements of Civil Engineering," published in 1839, names the authorities of that day who had devised methods for calculating the cross sections of bridge members.

Nevertheless, when the record sheets of their predecessors are totaled there remains no balance to subtract from the credit due to Squire Whipple and Herman Haupt. These men not only sifted correct methods from the mass of extant mathematical and mechanical data and organized a rational system; they also contributed original processes of their own, impelled by the current need for economy in bridge design.

1850—"DARK AGES" OF THE ROAD

1850 "DARK AGES" OF THE ROAD

Public roads were in a sad state of neglect, by 1850, when the railroads had captured the long-distance movement of passengers and freight in the populous areas east of the Allegheny mountains, as shown on the accompanying map. Although the aggregate operated railroad trackage totaled only 8,500 miles, as compared with an estimated sum of 71,000 miles of surfaced roads, the railways had gained such pre-eminence in public esteem that there was widespread apathy with respect to road improvement. The main turnpike roads were in a particularly vulnerable position. Long before the steam locomotive had demonstrated its supremacy, in 1830, the financial structures of many turnpike companies were tottering. The revenues derived from tolls on the infrequent and often massive horse-drawn vehicles were insufficient to meet the interest charges on the capital investment and leave enough margin to pay the operating and maintenance costs. Practically all the trunkline turnpikes, of which there was a large mileage, were in such financial straits, by 1830, that only the additional burden of railroad competition was needed to force their collapse. On the other hand there were many branch turnpike corporations acting as feeders to the railways, whose business was stimulated. This was true especially, after 1850, when the increasing horse-and-buggy traffic yielded unexpected revenues without serious destructive action upon the road surfaces. Thus, there was no marked slump in the number of new turnpike charters until after 1875. By that date the country had grown to such stature in population and wealth, that the original arguments for entrusting road improvement to private corporations, based upon the general poverty following the War of the American Revolution, had spent their force.

The stone-surfaced turnpikes, however, in 1850, represented only a fraction of the total network of public roads spread over the area between the Atlantic Coast and the frontiers in Western Missouri and Arkansas and eastern Texas, where there lived less than two persons to the square mile. Beyond were the long threads of transcontinental pioneer trails to the Far West in California and Oregon Territory. Generally speaking the eastern roads were stone covered for a few miles from the principal towns but beyond, the country thoroughfares were in a "state of nature". The railroads not only drove the freight-wagon and passenger-stage companies out of business but they also paralyzed any initiative to better the status of the local horse roads. The main highways, therefore, reverted to the jurisdiction of the county and townships authorities which had never lost control of the secondary routes. These public officials lacked the training, administrative experience, equipment and funds required to revive the public roads from the doldrums. Furthermore, road authorities were handicapped by the iniquitous maintenance system known as "statute labor," a relic of the feudal system in Europe. Isolated by bad roads for most of the year the predicament of the farmer, portrayed in the illustration, was a common sight.

RAILROADS IN OPERATION, 1850

1855~ISTHMUS OF PANAMA RAILROAD

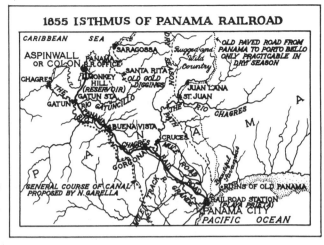

1855 ISTHMUS OF PANAMA RAILROAD

Improved communication between the Atlantic and Pacific Coasts became a matter of prime importance to the United States Government following the acquisition of California from Mexico and the settlement with England of the northern boundary dispute concerning Oregon Territory. During the war with Mexico, Benjamin A. Bidlack, United States chargé d'affaires at Bogota, signed a treaty with New Granada, on December 12, 1846, which granted to the government of the United States the right to cross the Isthmus of Panama by a wagon road, railroad or canal. A few months later, on March 3, 1847, Congress authorized the Secretary of the Navy to award mail contracts to steamships running between New York, New Orleans and Chagres, and from Panama to Oregon. The contract for the Pacific side was let to William H. Aspinwall of the Pacific Mail Steamship Company. The need for better communication to California became urgent, in 1848, when gold was discovered at Sutter's Mill near Sacramento.

The frenzied rush to the new gold diggings was fed not only by the eastern states but by people from all over the world. Thousands embarked on the long, monotonous sea voyage around Cape Horn at the southern tip of South America. Others took a steamer to the Isthmus of Panama where passengers and merchandise were conveyed in native bungos for a way and then transferred to the backs of mules which made painfully slow progress over the miserable roads to the Pacific side. There were other gold seekers who crossed the Isthmus by the more strenuous but less unhealthy Nicaragua route. Thousands of other treasure hunters followed the overland Oregon Trail from Independence, Missouri, through the South Pass to Salt Lake City where they branched off upon the direct road to California.

The walls of Congress echoed and re-echoed with orations describing the need for a Pacific Railroad or an Isthmian railroad, or canal, to accommodate the travel, trade and mail to the Pacific Coast which was growing in unprecedented volume. It was contended also that a Panama Railroad would give commercial superiority to the United States in the Pacific and California trade against our leading competitor, Great Britain. It was imperative also to complete adequate communications for the proper defense of California and Oregon in the event of war.

Thus after extended Congressional debate a Panama Railroad bill was enacted and engineers J.C. Trautwine and George M. Totten, established headquarters on the isthmus in January 1850, to begin the surveys. In spite of the tremendous construction difficulties which harassed the builders, mainly because of fever and malaria, seven miles of the railroad were opened to travel from Aspinwall (Colon) to Galinas in March 1852. Some three years later, on January 28, 1855, a locomotive ran from sea to sea over the 47½-mile track into Panama on the Pacific shores.

The superiority of the Panama Railroad was challenged by the establishment of the Butterfield Overland Mail on September 15, 1858, and ended with the junction of the Central and Union Pacific Railroads at Promontory Point, Utah, in 1869.

1856~MORMON HAND-CART EMIGRANTS

During the four years beginning with 1856 and ending with 1860 ten companies of Mormon hand-cart emigrants, shown in the accompanying illustration, crossed the Great Plains of the Far West to their Promised Land at Salt Lake City, Utah. In all there were a total of 662 hand carts and 2,969 emigrants. These Mormon hand-cart caravans constituted one of the most courageous and at times the saddest experiments in transportation recorded in the annals of western pioneers.

The hand cart, as the most economical vehicle, to aid Mormon converts in their 1400-mile journey from the Missouri River was devised by Brigham Young when the Perpetual Emigration Fund instituted to help needy converts became depleted. During the 1850's Mormon missionaries had won many proselytes to their faith in England, Wales, Denmark and other European countries. These people were clamoring for financial aid to make the trip to Utah. In answer to their pleas Brigham Young wrote a

1856 MORMON HAND-CART EMIGRANTS

letter on September 30, 1855, to F.D. Richards, President of the mission in Europe, in which he stated: "I have been thinking how we should operate another year. We cannot afford to purchase wagons and teams as in times past. I am consequently thrown back upon my old plan — to make hand-carts, and let the emigration foot it, and draw upon them the necessary supplies, having a cow or two for every ten."

Richards, in England, echoed these sentiments in an editorial in the church organ, the <u>Millenial</u> <u>Star</u>, dated March 1, 1856.

The first two hand-cart companies left Iowa City on June 9 and 11, in great spirits. They consisted of 497 persons and their belongings, 100 hand carts and 5 ox-drawn wagons carrying 25 tents and their groceries. Westward the train wound its way along the Platte River as groups among them throated the chorus of the hand-cart song: "Some must push and some must pull—As we go marching up the hill—As merrily on the way we go—Until we reach the valley, oh." They had their troubles but on the whole the journey was ended without serious mishap. The party crossed the South Pass in mid-September and were escorted into Salt Lake City on September 26 by welcoming members of their faith who met them at Emigration Canyon. A third caravan set out from Iowa City on June 23 and reached their destination tired but safely on October 2. Disaster, however, met the fourth and fifth companies which began their trek on July 15 and 26. Caught in the snow, 135 of the fourth group and 150 of the fifth lost their lives along the way before the caravans arrived at Salt Lake City on November 9 and 30, respectively. In 1857, only 480 hand-cart emigrants made the journey; in 1858, none; in 1859, one company; and the last in 1860 of two companies comprising 259 persons and 65 hand carts.

1857=THE CAMEL EXPRESS

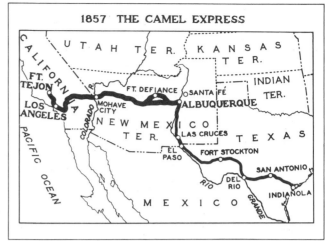

Our unique experiment in overland transportation, in 1857, intended to demonstrate the superiority of camels over horses, mules and oxen as beasts of burden, failed largely because the energetic Americans of the Far West were unfamiliar with the placid customs of the Far East. Camel transportation had been a favorite topic of speculation for some years when Senator Jefferson Davis, later President of the Southern Confederacy, amended an army appropriation bill during the last days of the 1851 Session of Congress. The amendment authorized an appropriation of $30,000 for the purchase and importation into the United States of 30 camels, 20 dromedaries, 10 Arab drivers and other necessary equipment. In advocating the amendment, Mr. Davis cited the deserts in Africa and Asia where camels were employed; stated that the English found them useful in the East Indies for carrying army supplies and light guns; and mentioned their value during Napoleon's Egyptian campaigns against natives bearing close resemblance to the American Comanches and Apaches. For these reasons, Mr. Davis believed that camels would be effective against our Western Indians. They could drink, before beginning a journey, enough water to last 100 miles; travel without rest at a rate of ten to fifteen miles an hour; and overtake Indian bands which eluded our horse cavalry. When another senator objected that camels were not worth more than $200 each and that our climate was too cold for their comfort, the amendment was lost.

The California newspapers then began a great hue and cry which finally overwhelmed the Congressional opposition. The California publicists advocated a "Lightning Dromedary Express" to carry fast mail and eastern newspapers to the Pacific Coast in 15 days. The camels were to fill their internal water tanks at the Missouri River, streak across the plains without regard for thirst, forage on sage brush along the way, drink deeply again at the Colorado River and amble into California coast towns two weeks from the time of starting. The camels were to be the solution of all western transportation problems until the Pacific railway was completed. Thus fortified with arguments, then Secretary of War Jefferson Davis, in the cabinet of President Franklin Pierce, succeeded, in 1854, in inducing Congress to appropriate $30,000 for the purchase and importation of the camels.

About 75 Bactrians (freight) and dromedaries (speed), purchased in Cairo, Egypt, and Smyrna, Arabia, were shipped to this country and unloaded from the naval-store ship "Supply", on February 10, 1857, at Indianola, Texas, on the Gulf of Mexico. Thence a group of the camels were despatched to Albuquerque, New Mexico, where an expedition was fitted out under the command of Lieutenant Edward F. Beale. The caravan followed the thirty-fifth latitude, approximating closely the present United States Route 66, and crossed the Mohave Desert to Fort Tejon, about 75 miles north of Los Angeles, California.

The camel experiment soon was admitted a failure by army officers. Every attempt to organize a camel caravan brought about incipient mutiny on the part of troopers and teamsters. Thus the "desert ships" were turned loose on the plains one by one. Some were killed by Indians who learned to enjoy camel steaks. Others died of neglect. At the outbreak of the Civil War, 35 or 40 survivors were herded into United States forts.

1858~BUTTERFIELD'S OVERLAND MAIL

The first through overland mail service between the Mississippi River and the Pacific Coast was provided by the "Butterfield Overland Mail" stage coaches, in 1858, as shown in the accompanying illustration. As early as 1849, concurrent with the gold rush to California, regular United States mails were transported by steamboats from New York City, carrying also express and passengers, to Chagres. Thence the trip was continued across the Isthmus of Panama by canoe and mule-back, until the railroad was completed in 1855, to Panama City where another line of steamboats completed the journey to San Francisco. The steamboats operated upon a semimonthly schedule. The trip from New York to the Golden Gate consumed 25 to 30 days.

During the nine-year period when steamboats supplied the only coast-to-coast transportation, local stage lines began running.

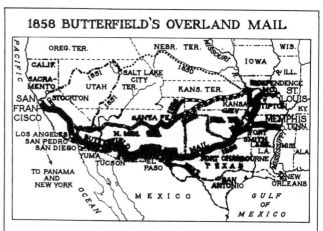

Supporters of the through overland mail service argued that only by this means could be broken the transportation monopolies of the steamship companies and the Panama Railroad. Overland railroads to the Pacific Coast were deemed far in the future. Furthermore, should a foreign war break out the enemy could readily sever the Panama mail route and divide the Union. Congress acknowledged the merits of these contentions by enacting the overland mail bill in March 1857. Postmaster General A.V. Brown, whose home was in Memphis, Tennessee, awarded the contract in September 1857, to John Butterfield, William G. Fargo and associates for a regular through mail service."

After a year of preparation the contractors launched the first stagecoaches simultaneously from St. Louis and San Francisco, on September 15, 1858, making the one-way trip in less than 24 days. Well-stocked relay stations punctuated the road at intervals of 10 to 15 miles, with armed guards in the Indian country. The "Butterfield Overland Mail" was a success from its inception. The sturdy Concord coaches carried nine people on three seats inside and one to ten more on top with the driver. By 1860 the amount of mail carried by the overland stages exceeded that transported over the Isthmus of Panama. During this period sectional interests were incensed over the choice of the southern "ox-bow" route because the location of the stage line was believed to fix the position of a future Pacific railroad. In response to regional pressure the Government opened a semimonthly mail line, in 1857, from San Antonio, Texas, to San Diego, California. In 1858, a mail service was authorized from Kansas City through Santa Fé to Stockton, California. In the same year a semimonthly mail was initiated across the Isthmus of Tehuantepec. Meanwhile the communications over the central route, from Independence through Salt Lake City to Sacramento, had been consolidated into a weekly operation.

The halls of Congress echoed with debates concerning the relative merits of the six mail routes to California. The debate was terminated in 1861 by the outbreak of the War Between the States and it became necessary, by July 1, to shift the overland mail line from the southern "ox-bow" to the "Central Route" where United States troops could preserve communications.

1858—BLAKE'S "STONE-BREAKER"

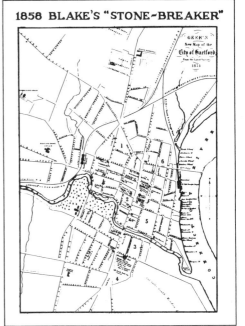

1858 BLAKE'S "STONE-BREAKER"

Blake's invention of an efficient "stone-breaker" was hailed with enthusiasm by "engineers, roadmakers, and lime burners." The machine was "strong, durable, and compact" and also portable. Depending upon the size of the opening of the jaws, the speed of the machine and the horse-powers of energy applied, the output of crushed stone varied from 4 to 6 cubic yards an hour. This quantity was very much greater than that produced by the slow, laborious method of breaking stone by hand with a rectangular or ball-shaped knapping hammer, as shown in the accompanying illustration. Perfected during the next fifty years, the power rock crusher and its indispensable adjunct, the power road roller, have multiplied many times over the miles of surfaced roads which could have been built in this country by manual labor.

Eli Whitney Blake, of New Haven, Connecticut, the nephew of Eli Whitney, the inventor of the cotton gin, patented his improved "stone-breaker" on June 15, 1858. Blake did not originate the idea of a mechanical rock crusher in this country. He was antedated by a "Machine for Breaking Stone for turnpiking and other purposes" patented on April 12, 1831, by Benjamin F. Lodge and Ezekial T. Cox, of Zanesville, Muskingum County, Ohio. A part of the description of their machine published in the Journal of the Franklin Institute follows: "A cast iron bed piece is to be provided upon which the stone to be broken is placed.***** A ram or hammer is to slide between cheeks, or to be hung like a tilt hammer. *****Any of the known motive powers may be used to work the hammer."

The inventors stated, "What we claim principally as new and useful, is the form of the bed, or block, and hammer, that is, the concavity of the one, and the convexity of the other; for we are aware that there has been an attempt made to introduce cast-iron plates, or blocks, with holes for the purpose of breaking stone as above; but having a level surface, and being worked by a common sledge, or hand hammer, they have been abandoned as useless."

It was his observation of the futility of crude mechanical crushing equipment similar to that described above that spurred Blake to produce a practicable "stone-breaker." He had had a good education. Graduated from Yale University in 1816, Blake immediately began the practice of law in Litchfield, Connecticut. For many years interested in public affairs, in 1851, he was a committee member selected to oversee the construction of about 1½ miles of macadam surface on Whalley Avenue, from New Haven to Westville, Connecticut. The hand method of breaking the stones with hammers caused the construction to drag along for two years. Blake tried to speed the work by having the rock broken upon an iron furnace grate with a trip hammer. When the process proved to be unsatisfactory he conceived the principle of his improved "stone-breaker." After his machine was built it was used first in Central Park, New York City, to produce the crushed stone for some concrete foundations. Blake's "stone-breaker" was employed first in highway improvement in Hartford, Connecticut. That municipality purchased two of his rock crushers in 1859 and 1860.

1859~MISSISSIPPI RIVER STEAMBOATS

The family tree of the romantic Mississippi River steamboats had its roots firmly established in the fertile soil of men's minds for at least 400 years. Steam as a means of mechanical power was known to the ancient Egyptians perhaps 6,000 years ago. Primitive steam engines are on record in succeeding ages. The first known reference, however, connecting a steam engine with the idea of propelling a ship was attributed, in 1543, to Blasco de Garay, a naval officer during the reign of Charles V of Spain. Nearly a century later, on January 21,1630, David Ramseye was awarded a patent by Charles I of England "to make shippes and barges go against strong wind and tide." A generation passed, then Edward Somerset, the second Marquis of Worcester, in 1651, spoke of using steam to drive the paddles of a boat. Not until 1707, however, was the first patent issued, for a boat actually built, to Denys Papin, a distinguished physician and scientist of Blois, France. James Watt began his experiments which resulted in a steam engine practicable for many purposes as well as propelling ships.

Inspired by the sight of Watt's engine, William Henry, an American inventor, returned from England in 1760 to his home in Lancaster, Pennsylvania. In 1763 he launched a steamboat for its trial trip on the Conestoga River near his residence. The boat sank, but he prophesied that, "The time will come when steamboats will be used to navigate the waters of the Ohio and Mississippi Rivers." Although his own venture failed, Henry managed to implant his vision on the minds of his protégés, John Fitch and Robert Fulton. Fulton's first attempt bore fruit, in 1785, when he operated a steamboat on the Delaware River at Davisville. In 1786 James Ramsey succeeded in propelling a crude steamboat on the Potomac River at Shepherdstown, West Virginia. Samuel Morey made a steamboat trial trip on the Connecticut River in 1790. Elijah Ormsbee drove a steamboat at a speed of 3 miles an hour on Narragansett Bay, Rhode Island, in 1792. Robert L. Stevens, son of Colonel John Stevens, crossed the Hudson River in his steamboat in 1804, from New York to Hoboken. Robert Fulton, in September 1807, made the celebrated trip of the Clermont.

Anxious to profit from the emigration to the west, Robert Fulton, associated with Chancellor Livingston and Nicholas J. Roosevelt, built the New Orleans at Pittsburgh, in 1811, the third steamboat to be launched on the Ohio River. Flat boats till then required 3 to 4 months for the upstream journey from New Orleans to Pittsburgh. Steamboats eventually reduced the time upstream to 21 and downstream to $10\frac{1}{2}$ days. Given a stimulus by the attacks on the New England coast during the War of 1812, emigration to the west was increased, in 1825, by the opening of the Erie Canal. By 1832 more than one-half of the western settlers journeyed in steamboats, according to Flint's History. The discovery of gold in California, in 1848, multiplied rapidly the steamboat business. By 1859, at the peak of the golden era, there were 2,000 steamboats on the big river and its tributaries.

1860 THE PONY EXPRESS

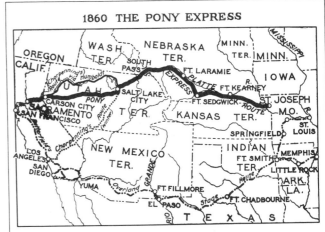

1860 THE PONY EXPRESS

The Pony Express was the first fast overland-mail service to the Pacific Coast. It brought San Francisco ten days closer to New York. During the critical days of the War Between the States, the Pony Express helped to preserve the Union by providing rapid communication between California and the seat of the Federal Government at Washington, D.C. Both the North and the South wanted California. For the Union, the State provided material resources and a base of operations against enemy aggression in the Far West. The control of California by the Confederacy could have stalemated the Federal Government west of the Rocky Mountains where half a million people lived.

There were three transcontinental mail routes to California before the Pony Express began operations. The first was the steamship voyage from New York to Chagres, thence across the Isthmus by canoe and mule to connect at Panama with the San Francisco steamship. These boats carried through the bulk of the western mail in 22 days. The second service, principally for local mail, ran over the Central Route from Independence, Missouri, along the Platte River, through South Pass, Salt Lake City, Carson City and Sacramento to San Francisco –total travel time 21 ½ days. The third, or Butterfield Overland Mail Company route, described an oxbow circuit some 600 miles south of the Central Route.

The plan for the Pony Express over the Central Route germinated from an idea outlined, in the fall of 1854, to United States Senator W.M. Gwin of California, by B.F. Ficklin, general superintendant of the western freighting firm of Russell, Majors and Waddell, in the course of a horseback trip from San Francisco to Washington, D.C. In January, 1855, soon after his arrival in the National Capital, Senator Gwin introduced in Congress a bill authorizing a weekly letter express, operating on a 10-day schedule between St. Louis and San Francisco. The bill was pigeon-holed in committee.

Senator Gwin revived the plan when the Civil War clouds began to gather, in 1860, by requesting Mr. Russell, then in Washington, to increase immediately the mail facilities of the Central Route in view of the probable closure of the southern road with the outbreak of hostilities. Mr. Russell hurried West, overcame the reluctance of his partners, and with characteristic Western energy, initiated arrangements for the Pony Express to begin 60 days later.

The Pony Express was opened officially on April 3, 1860, when riders left simultaneously from St. Joseph, Missouri, and Sacramento California, to race over the 1,966-mile distance. The first westbound trip was made in 9 days and 23 hours and the eastbound journey in 11 days and 12 hours-about one half the travel time of the Butterfield stage line. The pony riders covered 250 miles in a 24-hour day as compared with 100 to 125 miles by the stage coaches.

Although California placed reliance for news entirely upon the Pony Express during the exciting early days of the Civil War, the horse line was never a financial success. The Pony Express lasted only 19 months until October 24, 1861, when the completion of the Pacific Telegraph line, shown under construction in the accompanying illustration, ended the need for its existence.

1862—THE MULLAN ROAD

The 624-mile Mullan wagon road from Fort Benton, in Dakota Territory, the steamboat terminus on the Missouri River, to old Fort Walla Walla (Wallula), in Washington Territory, 127 miles easterly of The Dalles, the head of navigation on the Columbia River, was completed in 1862. The route followed for many miles the old trail traversed by the Indians on their way from the Pacific slope to hunt for buffalo east of the Rocky Mountains. Centuries of travel had worn this path in some places to a depth of twelve inches. The Mullan Road coincided very closely with the later Yellowstone Trail, now United States Route 10, across the State of Montana.

The improvement of this Indian trail to the status of a wagon road was authorized by Congress early in 1857, implemented by an appropriation totaling $100,000 placed at the disposal of Secretary J. Holt of the War Department. The reconnaissance survey was made in 1858 by Lieutenant John Mullan, Second Artillery Regiment, United States Army, for whom the road was named.

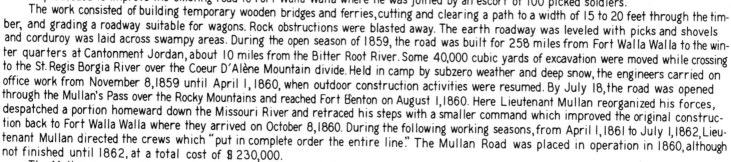

1862—THE MULLAN ROAD

Arriving at The Dalles, Oregon State, on May 10, 1859, as directed by Captain A. A. Humphreys, United States Topographical Engineers, Lieutenant Mullan proceeded to improve the existing road to Fort Walla Walla where he was joined by an escort of 100 picked soldiers.

The work consisted of building temporary wooden bridges and ferries, cutting and clearing a path to a width of 15 to 20 feet through the timber, and grading a roadway suitable for wagons. Rock obstructions were blasted away. The earth roadway was leveled with picks and shovels and corduroy was laid across swampy areas. During the open season of 1859, the road was built for 258 miles from Fort Walla Walla to the winter quarters at Cantonment Jordan, about 10 miles from the Bitter Root River. Some 40,000 cubic yards of excavation were moved while crossing to the St. Regis Borgia River over the Coeur D'Alène Mountain divide. Held in camp by subzero weather and deep snow, the engineers carried on office work from November 8, 1859 until April 1, 1860, when outdoor construction activities were resumed. By July 18, the road was opened through the Mullan's Pass over the Rocky Mountains and reached Fort Benton on August 1, 1860. Here Lieutenant Mullan reorganized his forces, despatched a portion homeward down the Missouri River and retraced his steps with a smaller command which improved the original construction back to Fort Walla Walla where they arrived on October 8, 1860. During the following working seasons, from April 1, 1861 to July 1, 1862, Lieutenant Mullan directed the crews which "put in complete order the entire line." The Mullan Road was placed in operation in 1860, although not finished until 1862, at a total cost of $230,000.

The Mullan wagon road was projected to transport supplies to forts, camps and homes, between the head of navigation of the Missouri and the Columbia Rivers, not easily accessible from the Oregon Trail to the South. The Mullan wagon road relieved the freight congestion over the Oregon Trail, shortened the distance to be traveled by wagons, lessened the hardships of emigrants because of the longer boat journey and a-
voided the Indian raids along the Sweetwater and North Platte Rivers.

1864—THE BOZEMAN TRAIL

The Bozeman Trail has been called the Montana Road, the Jacobs-Bozeman Cut-off, the Powder River Road to Montana, the Big Horn Road, the Virginia City Road, the Bonanza Trail, the Yellowstone Road, the Reno Road and the Carrington Road. The 967-mile Bozeman Trail traversed the tribal lands of the fighting Sioux, the Cheyennes, the Arapahoes and the Blackfoot Indians, a confederation led by the great Ogallala Sioux war chief Red Cloud. The Bozeman Trail provided a short cut from the Oregon Trail to the new gold diggings at Virginia City, which became the capital of the territory of Montana, in 1864, with a population of 10,000 fortune hunters, after the precious metal was discovered in Beaverhead Valley. The Bozeman Trail, during the four years of its brief existence, supplanted the earlier roundabout routes to the gold-strike region: (1) By way of the Missouri River to Fort Benton and overland; and (2) along the Oregon and Overland Trails to Fort Hall, thence northerly over the Virginia City road.

The Bozeman Trail began at Fort Kearney, Nebraska, where all soldiers and equipment were assembled for battling the Indians of the region. For all practical purposes, however, the starting point was at Fort Sedgwick.

The Bozeman Trail was named after John M. Bozeman, an adventurous Southerner from Coweta County, Georgia, who, with John M. Jacobs, in the winter of 1862-63, determined to find a shorter route to the Montana gold fields for emigrants and freighters. Enduring severe hardships and risking their lives in the unfriendly Indian country, Bozeman succeeded, in 1864, in piloting a long train of emigrant wagons over his chosen path. The Bozeman Trail traversed the heart of the Indian lands forbidden to the white man by a treaty made, in 1851, between the United States Government and the Sioux nation. When, in disregard of the treaty, white men sought to possess the region many fierce battles ensued.

After continual frustration and some defeats, the Government forces won the decisive Wagon Box victory, in 1867, where 27 United States soldiers and 5 teamsters, armed with the new breech-loading rifles, routed some 3,000 Indians with the loss of nearly half their number. True to his word, Red Cloud never fought again but Sitting Bull and Crazy Horse continued the struggle nine years later in the 1876 Sioux War when Lieutenant Colonel George A. Custer made his last stand at the Little Big Horn River.

Considering it expedient to seek a treaty, in 1868, the Government was forced to accept Red Cloud's terms providing for the abandonment of all the forts erected throughout the Indian hunting grounds and the discontinuance of further efforts to open a wagon road into Montana. The entire region thus reverted to the possession of the victorious Sioux nation.

The following epitaph is inscribed on a stone monument erected in the peaceful cemetery crowning the brow of the bluff overlooking the beautiful city of Bozeman, Montana, "In memory of John M. Bozeman, aged 32 years, killed by Blackfoot Indians on the Yellowstone, April 18, 1867. He was a native of Georgia, and was one of the first settlers of Bozeman, from whom the town takes its name."

1866 – THE "BONESHAKER"

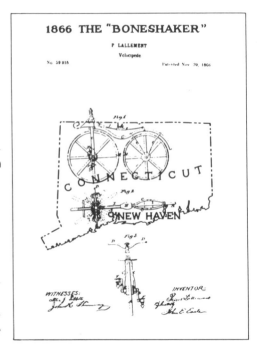

The crank-driven velocipede, the prototype of the modern bicycle, introduced in France, became known as the "boneshaker" because of its rough-riding characteristics. This machine was the next important link in the long chain of development following the "hobby-horse" of 1819, which was soon relegated to oblivion in England and America mainly because of its clumsy design. For nearly two decades thereafter velocipedes attracted only passing attention until interest was revived in 1834 by Kirkpatrick Mc Millan, a blacksmith of Keir, Dumfriesshire, Scotland. This ingenious pioneer is said to have fitted pedals with connecting rods to the rear axle of a tricycle and later, in 1840, to have added cranks, pedals and driving rods to a "hobby-horse." Mc Millan also equipped the velocipede with a comfortable seat, a padded arm rest and a handle bar. He traveled about the countryside so fast that he was arrested once and fined for "furious driving." The canny Scotsman seems to have been the first to apply motive power to the rear wheel. About 1845 another Scotsman, a cooper by trade, named Gavin Dalzell, living in Lesmahagow, Lanarkshire, improved Mc Millan's velocipede. The vehicle was fitted with wooden wheels shod with iron tires. The front wheel was about 30 inches and the rear 40 inches in diameter. Following these innovations of the two Scotsmen, two decades elapsed before another important contribution was made toward the improvement of the velocipede.

Then, in 1865, rotary cranks were added to the front wheel of the French velocipede. Credit for the invention has been given to two men. One was Ernest Michaux, the son of a Parisian carriage maker, to whom a monument was erected, in 1894, in Bar-le-Duc, near Verdun, France, for being the first person to suggest cranks, in 1861 or 1863. The other was Pierre Lallement, a blacksmith in the employ of M. Michaux, whose friends claimed him the inventor in 1865. Lallement became dissatisfied with the treatment accorded him at the plant, sold the patent to his employer and emigrated to the United States in 1866, where he made a velocipede and rode it about the streets of New Haven, Connecticut.

In France, however, the Michaux Company had continued to make improvements on Lallement's patent so that in 1867 the new vehicle became the sensation of Paris. Riding schools hung out their signs throughout the city. Straps for holding the machine became standard equipment at places of amusement. The news spread across the Channel to England. There, in 1868, the Coventry Sewing Machine Company expanded their manufacturing plant under the firm name of Coventry Machinists' Co., Ltd., in order to handle an order for 500 velocipedes placed with them by their Paris agent. The Franco-Prussian War broke out when only a few of the machines had been delivered. To avoid financial loss the company unloaded the remainder upon their home market. This circumstance established the trade in England. Meanwhile, in 1866, Edward Gilman, of England, had patented a rear-driven bicycle with a single pedal. Also there had been intimations about a chain gear.

Word about the improved Michaux velocipede was voiced in the United States where it came suddenly into vogue in 1869.

1866–DUDGEON'S STEAM CARRIAGE

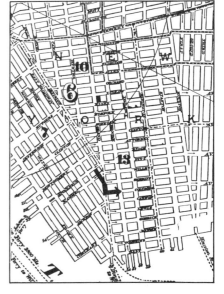

1866 DUDGEON'S STEAM CARRIAGE

In a catalogue advertising hydraulic jacks which he had invented, Richard Dudgeon wrote, in 1870, "Above I have given a good wood engraving of my last steam carriage, as a number have expressed interest or curiosity in it.

"This is not in the way of business or advertising at all. After seventeen years of effort and conviction of its utility, I have learned that it is not fashionable, or that people are not ready for it. *****

"Without any patents about it, it will go all day on any good wagon road, carrying ten people at fourteen miles an hour, with 70 lbs. of steam, the pump on the fire door open, if desired. One barrel of anthracite coal is required to run at this speed for four hours. It weighs 3,700 lbs. with water and fire to run one hour. It will go 20 miles in an hour on a good road. It is perfectly manageable in the most crowded streets."

In spite of all that could be vouchsafed in its favor, however, the disillusioned inventor of the steam carriage, shown in the accompanying illustration, at last had become reconciled to the fact that the public was not interested. In 1855, Dudgeon had astounded New Yorkers by riding from his home on East Broadway to his place of business at 24 Columbia Street in the first steam carriage which he had assembled. It was built as the result of a wager between himself, a determined Scotsman, and two of his associates. The "Red Devil Steamer" aroused immediate newspaper criticism, "The running of the wagon is accompanied by a great deal of vibration and noise, for there are four exhausts, as in a locomotive, and the solid wooden discs that serve for wheels pound the road heavily." The commotion frightened horses so badly that the irate drivers appealed to the city authorities who issued a city permit limiting the operation of the steam carriage to one city street only. A few years later the machine, while on display in the Crystal Palace, was lost in the fire which razed that imposing edifice. Sensing keenly the need for mechanical road transport in an industrial age, Dudgeon constructed his second steam carriage in 1866. Again he encountered the outspoken opposition and derision of drivers and riders of horse-drawn vehicles. Restricted as before in its range on city streets, Dudgeon moved his family and steam carriage to Long Island. There the vehicle became a familiar sight with a negro boy running ahead to warn travelers of the danger that followed. Dudgeon ran the vehicle many hundreds of miles. The longest trip was from Locust Valley, Long Island, to Bridgeport, Connecticut, and the fastest speed was a mile covered in one minute and fifty-two seconds. Dudgeon's device, however, failed to gain popular favor. It was a brain child born out of due time.

Dudgeon's failure followed a repetition of the popular apathy encountered by Thomas Blanchard, who manufactured a steam road car in 1825. The machine ran along the streets of Springfield, Massachusetts, proved to be a sturdy hill climber, was endorsed by the State Legislature, was gaped at from the sidewalks, some few intrepid souls took a ride but the public at large would have none of such a strange contraption.

1869~THE MEETING OF THE RAILS

1869 THE MEETING OF THE RAILS

The Last Spike, driven on May 10,1869, at Promontory Point jutting southward into Great Salt Lake 22 miles west of Ogden, Utah, joined the Central Pacific and Union Pacific railroads into the first transcontinental railway in this country. This event climaxed a generation of struggle to speed overland communication following the discovery of California's gold in 1848. The Federal Government had extended a helping hand as early as 1849 when the first United States mail service to the Pacific Coast began moving by canoe and muleback across the Isthmus of Panama. Steamships carried the mail from New York to Chagres and from Panama to San Francisco making the coast-to-coast trip in the fast time of 25 to 30 days.

While these mail steamships were serving the California pioneers, a light wagon, or packhorse, overland United States mail line began traveling, in 1850, from Independence, Missouri, to Salt Lake City, Utah. The round trip lasted sixty days. A mail route was extended westward from the Mormon capital to California, in 1851. Because a railway had been opened across the Isthmus, in 1855, the Panama route provided the most rapid and convenient transportation to the West Coast until 1858.

Californians, however, wanted a faster overland service and the halls of Congress reverberated with the wordy artillery of their representatives. Their bombardment brought victory in March, 1857, when the overland-mail bill was enacted. John Butterfield, William Fargo and associates were awarded a $600,000 one-year contract, during the following September, for a semi-weekly overland mail service. The chosen route began at St. Louis and Memphis, converged at Fort Smith, Arkansas, thence passed through Tucson, Yuma and Los Angeles to San Francisco, a total distance of 2,795 miles from St. Louis. Starting from both ends of the line, the first coaches made the trip in less than 24 days.

Nevertheless, the aggressive Californians pressed for still faster communications with the National Capital.

Following the outbreak of the War Between the States, the equipment of the Butterfield Overland Mail was shifted north to the Central Route, on July 1, 1861, so as to gallop across Union-controlled territory with its daily overland mail between St. Joseph and San Francisco.

The railroad now heeded California's plea for speedier western communication. On July 1,1861, President Abraham Lincoln signed the Pacific Railroad Act designating two railroad companies to build and operate a combined railroad and telegraph line between the Missouri River and Sacramento. The Central Pacific Railroad started construction on the California end on January 8,1863. The Union Pacific Railroad turned the first shovelful of earth at Omaha, Nebraska, on the following December 2.

In the accompanying photograph, are shown the junction of the rail lines, the Concord coaches, so named because manufactured at Concord, New Hampshire, and the Chinese coolies who made the construction of the railroad possible during a period of labor scarcity.

1869~ THE STEAM ROAD ROLLER

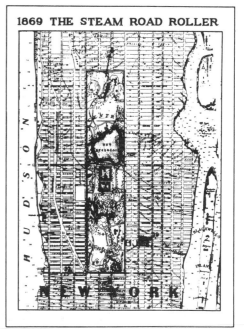

1869 THE STEAM ROAD ROLLER

The report concerning the first steam road roller used in the United States, shown in the accompanying illustration, was published in the June 19, 1869 issue of the Scientific American Magazine as follows: "A trial of the new steam road roller, purchased by the Central Park Commissioners to be used on the roads under their charge was made on June 4th at the corner of 115th Street, and 6th Avenue in this city.

"The machine was made by Aveling and Porter, Rochester, England, and we are informed, weighs about fifteen tuns. It has four rollers, two front, and two back, so placed that the hinder ones cover the ground not rolled by the front ones.

"Two of the rollers, perform the office of drivers; being turned by an endless chain and rag wheel; the others are made to turn like the forewheels of a waggon to guide the machine. The engine runs with a quick stroke and is speeded down so that great tractive power is obtained.

"The ground on which the machine was exhibited, was of a very friable kind, being composed mostly of a coarse sand. We think its operation would have been still more satisfactory than it was, had the character of the ground been different. As it was, we believe all present were satisfied of the great efficiency of the machine, though we heard some improvements suggested. These were however made too hastily to be perhaps of much value.

"We understand that this roller, has been used largely as a traction engine for moving heavy weights in the iron-works of London, and it seems admirably adapted to that purpose."

This steam road roller was tested thoroughly in England before being shipped to this country, according to the Journal of the Franklin Institute which contained the statement in April, 1869, that, "A public trial was made at Rochester, on Friday, January 15, in the presence of a number of officers of Royal Engineers and others, of a powerful steam road roller recently manufactured by Messrs. Aveling & Porter, at their works, at Strood, and intended to be sent out to New York, to be used on the roads of the Central Park in that city. The engine was purposely tested under the most disadvantageous circumstances, with a view of fully developing its power, for which purpose it was made to ascend Star-hill, the steepest incline in the city, which has a rise of one in twelve $\overline{/8\frac{1}{3}}$ per cent grade/ The entire surface of the roadway had previously been thickly covered with stones of the ordinary kinds used on macadamized roads. The steam roller commenced its work soon after 10 o'clock, and notwithstanding the increased difficulties it had to surmount, by 4 o'clock in the afternoon, it had made repeated ascents and descents of the hill, the entire surface of which was rolled completely smooth and fit for the passage over it of the lightest vehicles."

The history of road rolling in England may be traced back to the patent granted, in 1619, to John Shotbolte for employing "land stearnes, scowrers, trundlers, and other strong and massy engines ***** in the making and repairing of high way and roads."

115th St

1871~ THE CHISHOLM CATTLE TRAIL

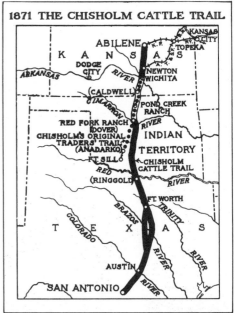

1871 THE CHISHOLM CATTLE TRAIL

A broad, beaten thoroughfare, several hundred yards in width, "tramped hard and solid by the millions of hoofs which had gone over it," the Chisholm cattle trail traversed part of the 800-mile distance northerly from San Antonio, Texas, to Abilene, Kansas, according to Sam P. Ridings. Rated the first place among all American cattle paths, the Chisholm trail was the outlet to the northern markets for the three million long-horn steers and other cattle which had multiplied in Texas while commercial intercourse was halted by the long and costly War Between the States. The longhorn steers, descended from the animals the Moors had driven into Spain and the Spaniards had introduced to America, were so named because their bony weapons standing out at right angles from their heads measured often 7 feet or more from tip to tip.

Although the southern extension of the trail is spoken of ordinarily as running northeasterly from San Antonio to Austin, actually the southern terminus was situated at the ford across the Colorado River three miles downstream from Austin. To that crossing cattle converged over many routes leading from the regions of southern Texas and the Gulf Coast. From Austin the cowboys drove the herds north over the Brazos and Trinity Rivers to the crossing over the Red River north of the present city of Ringgold, Texas. Thence the drive continued parallel and east of the existing United States Route 81 across the State of Oklahoma to Abilene, Kansas, the first "Cowboy Capital of the West."

The original Chisholm trail was opened for traders' wagons, in 1865, by Jesse Chisholm, a famous half-breed Cherokee plainsman and scout who had erected a trading post on Chisholm Creek, on the site of Wichita, Kansas. For 150 miles between Wichita and Dover the cattle trail followed the 220-mile traders' path extending to near present Anadarko, Oklahoma. The section of the trail immediately to the south, between the Cimarron and the Red Rivers, was broken for the passage of the first herd of cattle, in 1867, by Colonel O. W. Wheeler, a resourceful cattleman. In that year 35,000 cattle were shipped by rail from Abilene; during the next year 1868, 75,000; in 1869, 350,000; in 1870, 300,000; with an all-time high for this rail head of 600,000 in 1871. The life of the trail began to ebb at noon on April 22,1889, the time set in the proclamation of President Benjamin Harrison opening to settlement the Government-owned lands in the new Territory of Oklahoma.

Dodge City further west won preeminence as the "Cowboy Capital of the World."

During the long, tragic years of the Civil War there had survived no northern market for Texas cattle. Outlet to the east was blocked for the greater portion of the struggle by northern armies on the Mississippi River. At the conclusion of hostilities the extensive unexplored lands lying between the Red and Cimarron Rivers continued to be a barrier to the movement of cattle. The eastern portion of this wilderness had been established as Indian Territory, in 1832, to provide a haven for the Choctaw, Cherokee, Chickasaw, Creek and Seminole tribes deported from their home lands.

1872~THE "ORDINARY" BICYCLE

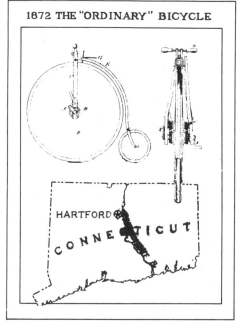

1872 THE "ORDINARY" BICYCLE

After the "boneshaker" was relegated to the scrap heap, in 1870, Americans lent scant interest to velocipede locomotion for nearly a decade. In the meantime there had continued a steady betterment in England in the quality of the "boneshaker." This improvement was stimulated by their hard, smooth macadam roads with attractive side paths distant far enough so that velocipedists did not frighten horses. The new machine was patented in England on April 8, 1869, as a bicycle—the first time that this name seems to have been given official recognition. This bicycle was equipped with steel rims surmounted by solid rubber tires which minimized vibration and reduced the amount of power required for propulsion over road surfaces. As a result the size of the front wheel was increased subsequently to gain speed without exerting any more muscular effort than that required by the clumsy "boneshaker." By 1871 the diameter of the front wheel had been increased to 40 and 48 inches and the rear wheel reduced to 16 inches — a size just sufficient for steering purposes. J. K. Starley, of England, in 1872, introduced his "spider-wheel" contrivance, one of the earliest of the fully-developed "ordinary" bicycles. The "high bicycle" with front wheels which reached a maximum diameter of 60 and 64 inches was extremely popular in England from 1872 to 1885, when many other forms of pedomotors were being experimented with such as monocycles, tricycles, four- and five-wheelers.

The popularity of the "ordinary" bicycle was destined to be short-lived because of basic defects in design, the most serious of which was the placement of the center of gravity of the rider almost over the center of the front wheel. The net result was a condition of instability.

For these reasons the ingenuity of inventors sought to devise a new pattern of bicycle by exploring in apparently opposite directions. In the first case, the size of the front wheel was maintained nearly constant but the saddle of the rider was moved further to the rear by the introduction of levers, cranks or chain gears between the pedals and the main driving axle. In the second case, the rear wheel was made larger than the front.

"Ordinary" bicycles did not make their appearance in the United States until 1876, when several English makes were displayed at the Centennial Exposition commemorating our independence held at Philadelphia, Pennsylvania. There, as he eyed the new velocipedes, Colonel Albert A. Pope, of Boston, Massachusetts, became fired with the desire to revive the wheel industry in this country. He sailed for England soon after and returned with samples of their best "ordinary" bicycles. Colonel Pope prevailed upon W. S. Atwell of Boston, to build a duplicate which weighed 70 pounds and cost $313. Visiting England once more to confirm his initial judgment, Colonel Pope returned home determined to begin manufacture of the improved bicycle at once. As a beginning he entered into an agreement with the Weed Sewing Machine Company of Hartford, Connecticut, in 1878, to construct some "Columbia" "ordinary" bicycles on some unused floor space in their plant. These were the first bicycles of the "ordinary" type made in the United States for sale to the public. The accompanying illustration shows "ordinary" bicycles on Pennsylvania Avenue in Washington, D.C., during the 1884 meeting of the League of American Wheelmen.

1880-WORKING OUT THE ROAD TAX

When the League of American Wheelmen, organized in 1880, took up their cudgels to battle for Good Roads over the length and breadth of the country, the wasteful statute-labor system was in vogue everywhere. This extravagant method of road repair under the jurisdiction of local county and township authorities was known also as "working out the road tax." The practice was legalized during the Middle Ages in England and was patterned after the customs of the Roman empire.

Alfred Leger related, in his work describing Roman public works, "The parish roads were constructed at the expense of the towns and boroughs, with the assistance of some special donations, and supported by the oath for three days labor made by the adjoining property owners."

Following the dissolution of the Roman empire in the West in 476 A.D., systematic road work was abandoned. Some centuries later, however, the Roman methods were restored in part by the Frankish Kings. The statute-labor system, called the corvée, became firmly embedded in the administration of the feudal states which the Frankish empire erected during the sixth to the tenth centuries upon the remains of the Gallo-Roman estates. The corvée survived throughout the Middle Ages and well into the nineteenth century in England and into the first third of the twentieth century in the United States. The word corvée is a corruption of the Roman operae corrogatae, that is requisitioned works derived from rogare, to request, and shortened to corvatae, corveiae and lastly to corvée.

During feudal times the barons assumed the sovereign power formerly exercised by the central authority of the state. Corvées represented serf labor in part for payment of rent to the feudal lord and in part to meet the taxes imposed for the repair and building of roads, castles and churches and the carrying of letters and dispatches. The corvée royale, or unpaid road labor, made a general obligation in France in 1738, became so burdensome to the peasants that it was one of the most potent influences contributing to the French Revolution.

Statute labor was introduced from England into the British Colonies in North America. The costly character of the system was recognized by leaders in this country from the earliest times. George Washington inveighed against it in a letter to Patrick Henry, dated November 30, 1785, in which he advocated the abandonment of county-controlled statute labor and the substitution of the contract work directed by a central authority. Governor Livingston, of the State of Pennsylvania wrote in the November 1791, issue of The American Museum, "Roads ought, in my opinion, to be repaired at the expense of the county, and for the money raised by such tax, men might easily be found, that would work, and expect to work faithfully, instead of the ridiculous frolic of a number of idlers, in which that important business is at present converted.***** "

The statute labor system, shown in the accompanying illustration, survived in this country until the pressure for better roads generated by the bicycle and the automobile outweighed the resistance of those who benefited from the payment of taxes in work instead of cash.

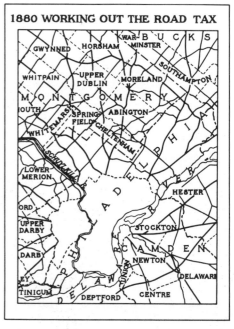

1880 WORKING OUT THE ROAD TAX

1885 – THE "SAFETY" BICYCLE

The basic principle of the "safety" bicycle was discovered by Harry J. Lawson, of London, England, son of a prominent Brighton preacher. Lawson's objective was to make the rider "safe" by doing away with the "headers" over the handlebars of the high "ordinary" bicycles, the cause of accidents which had terminated in so many fractured skulls and bruised bodies. Lawton's father is said to have told him that "headers" would be impossible if the center of gravity of the rider were made lower than the center of the driving wheel. This relative position could be attained only by redesigning the old "high bicycle" so that the pedals were connected to the main axle by levers or other mechanical device. In the course of perfecting his design, begun in 1868, Lawson shifted the drive to the rear wheel as the most practical solution of the problem. He patented his rear-wheel driven "safety," with a lowered front wheel, in 1876. When Lawson arrived in Coventry in 1877, prepared to demonstrate his patent, he met with a cool reception from the "ordinary" bicycle manufacturers. These practical business-men were making handsome profits on the popular "high bicycle," and they could see no good reason for introducing another model which changed the driving power from the lowered front to the rear wheel by the addition of a sprocket-wheel and chain attached to a third axle. Ideas, however, never die. They only sleep ready to spring to life whenever conditions are ripe to give them substance.

Other bicycle makers, including Humber and Company, Singer and Company, and Starley and Sutton, devoted part of their energies toward perfecting Lawson's "safety." Starley won the race in 1884 by designing the famous "Rover" rear-drive bicycle which established the pattern for the modern "safety." The front wheel was one-fourth larger in diameter than the rear wheel over which was located the saddle. Later the wheels were made nearly equal in size. Commercial success for the new "safety" was assured by the widespread sales following the advertising campaign launched by Starley and Sutton, in 1885.

At this time the "ordinary" bicycle in the United States was all the rage. Intimations from abroad about the superiority of the "safety" were ignored or received with skepticism until after Dr. John Boyd Dunlop, a veterinary surgeon of Belfast, Ireland, patented his pneumatic tire in 1888–1889. This innovation was the most important single improvement during the many centuries of bicycle development. The pneumatic tire made the "safety" so comfortable to ride and so speedy that by 1890 the "ordinary" bicycle in the United States was considered a "back number." Dr. Dunlop conceived the idea after fitting the wheel of his son's bicycle with a rubber hose to cushion the vibration. His invention was an improvement over the preceding solid-rubber tires which had been added to the former solid-iron tires. Dr. Dunlop's patent revived the original invention of the pneumatic tire by R. W. Thomson, an English civil engineer, who had been awarded a patent in 1845 which later lapsed without commercial exploitation.

1889 ~ SAND ~ CLAY ROADS

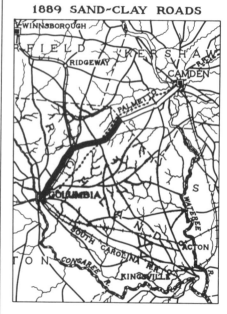

1889 SAND~CLAY ROADS

Sand and clay combined in the proportions best suited to provide a stable road surface have been found at many locations on the surface of the globe since time immemorial. Taking their cue from Nature, roadbuilders of the ancient and modern worlds have added clay to sandy roads to give them stability or added sand to clay surfaces to prevent them from rutting and becoming sticky in wet weather. The relative proportions required to yield a dense, compact surface varied with the character and grain sizes of the materials. Generally speaking, however, the amount of clay should not exceed the volume of voids in the sand. Thus the clay acted as a binder while the sand particles were permitted to bear upon each other and interlock so as to resist displacement by traffic. This type of construction is limited to those areas where the frost does not penetrate the ground to any appreciable depth.

References to the construction of mixed sand and clay surfaces may be found in the earliest roadbuilding literature in this country. S. W. Johnson, in 1806, in his "Rural Economy Containing a Treatise on Pisé Building" describes the use of a sand subbase over clay soils. W. M. Gillespie wrote in his manual in 1852, "If a soil be a loose sand, a coating of six inches of clay carted upon it will be the most effective and the cheapest way of improving it, if the clay can be obtained within a moderate distance. Only one-half the width need be covered with the clay, thus forming a road for the summer travel, leaving the other sandy portion untouched, to serve for the travel in the rainy season.

"If the soil be an adhesive clay, the application of sand in a similar manner will produce equally beneficial results."

Sand-clay surfacing, because it was lower in cost, adequate for light traffic and less dusty and noisy and more resilient than macadam, was the logical answer to the road problem of the South Atlantic and Gulf States, in 1885, when the League of American Wheelmen were reviving interest in Good Roads. With their economy depleted by the ravages of war, the South needed a road surface that could be built and maintained at a small cost from local materials in generous abundance. It was a logical development, therefore, that S. H. Owens, road supervisor of Richland County, South Carolina, because of his location and the thoroughness and efficiency of his work, should be acclaimed by his associates as "the father of the sand-clay road in America."

Mr. Owens wrote, "In January, 1889, I took charge of the roads in this county, which were then in deep sand in two-thirds of the county, the balance being through sticky clay hills with the exception of about two miles of macadam road which had proven too expensive for our county to continue to build."

"I commenced covering the sand on the old Camden road with clay, to about 10 inches in depth. At first the people were displeased. It had rained a great deal and they were not accustomed to seeing muddy roads. I continued to throw sand on the clay until it quit bogging and sticking to the wheels, keeping it crowned with an ordinary road scraper. After I had built a few miles of the road and it became smooth and hard the people were delighted."

1892—BICYCLING DAYS

The pneumatic-tired "safety" bicycle had captured the imagination of the American people to such an extent by 1892 that newspapers and periodicals of that day referred to the bicycle "craze." The mania found expression in riding schools organized to instruct novices in all the large cities; in the "Daisy Bell" popular song featuring "A Bicycle Built for Two"—the tandem; by slang words as "scorching", an exaggeration of the heat produced by the friction of the speeding vehicle with the road surface; and by the novel "rational dress" called "bloomers" which the ladies adopted so as to acquire greater freedom of movement. This costume had been contrived, in 1851, by Mrs. Amelia Jenks Bloomer, an ardent advocate of women's rights and dress reform. Lacking any real need for its adoption, her innovation was forgotten until revived by the bicycle. The "safety" left in its wake not only a new wave of styles in women's and men's dress but also a new start in the improvement of the roads and streets of this country.

In the "Gay Nineties" from 1890 to 1895, when the "safety" was being standardized and perfected, only a small mileage of city streets were surfaced with pavement, macadam or cobblestones. The country roads as a whole were in a miserable condition. The old macadam turnpikes built during the early part of the nineteenth century had been allowed to deteriorate after 1835 to 1850 when superseded progressively from east to west by the gleaming steam railroad tracks. The League of American Wheelmen, organized in 1880 by a consolidation of local "ordinary" bicycle clubs, had been sounding a clarion call for GOOD ROADS and bicycle sidepaths for more than a decade. Their monumental efforts began to bear fruit as State after State enacted local road-aid laws, led by New Jersey in 1891. At that time the plight of the "city slicker" cyclist caught in a thunder shower on a hilly earth road was a source of amusement to many a farmer, as suggested in the accompanying illustration. Road signs were either missing or illegible, and maps presented meager and often inaccurate information. This picture illustrates the axiom of highway development that the improvement of the roadway always lags behind the evolution of the vehicle.

During the period from 1892 to 1894, the bicycle manufacturers enjoyed great prosperity, boosted further by exports to England. The decline in demand began in 1897 for reasons that seem obscure. The competition of the effortless automobile does not explain satisfactorily the abrupt cooling of popular enthusiasm. One cause that has been suggested is the slowing of demand resulting from the change in policy of manufacturing new models every three or four seasons instead of annually as before. Perhaps the older generation found bicycling a more strenuous exercise than they anticipated even though the "safety" weight totaled only 9 to 35 pounds, as compared with 19 to 60 pounds for the "high bicycle." Whatever the cause, many more bicycles are ridden today than during the halcyon years of the "Nineties." In 1896 there were 4 million bicycles in use.

73

1892—FIRST STATE-AID ROAD NEW JERSEY

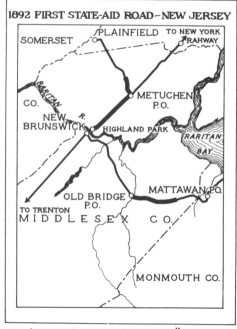

The unique feature of the State-aid road law which originated in New Jersey consisted in the "granting of direct aid by the State to the permanent improvement of its highways." The law was approved first by the Governor of the State on April 14, 1891 but remained inoperative until amended on March 29, 1892, to impose "the work of inaugurating and enforcing the law" ***** "upon the President of the State Board of Agriculture." Supplementary legislation abolished the overseer method of road improvement established by the general road law of 1846. The State-aid road law provided for the appointment of Township Committees who "should annually inspect the roads of their townships, and adopt a systematic plan for improving the highways; they have power to employ an engineer or any competent person for advice, plans and estimates; they can select and appoint one or more persons to do certain kinds of work, and others for another kind; they can give a taxpayer an opportunity to work the amount of his tax out, or any portion of it, by paying him the cash for his work or accepting his bill, but in all cases they should say when, where and how the work must be done, and make it conform to their system of improving roads" ***** "It is estimated that at least 27,000 tons of water fall annually on one mile of road, and the necessity of a well-rounded road-bed, with open side ditches from outlet to outlet, is an important feature."

The State-aid financial principle of the law provided that the State government pay one-third of the total cost of construction of a road project, the adjoining property owners one-tenth and the local county the remainder plus the future maintenance of the work. "A person desiring to have a road improved must first determine that the road is not less than one mile in length, and that it does not approach a county or municipal line."

"The passage of this law created a demand by the friends of the road movement for its enforcement, and an equally determined opposition, which resulted in an appeal to the courts, and the mandatory features were sustained. During this opposition elsewhere the county of Middlesex, seeing an opportunity to get several roads built immediately, borrowed $50,000 or $60,000 for road purposes for their share of the expense, and extended the proposed roads to be improved five miles more or thereabouts, making about ten miles to be improved under the new law, and the State paid them on the 27th of December, 1892, $20,661.85, being the first money paid by the State of New Jersey for improved roadways."

The work consisted of a total of 10.55 miles of macadam construction divided into 4.85 miles between New Brunswick and Metuchen, twelve feet wide and eight inches thick; 1.70 miles being part of the road between Metuchen and Plainfield, fourteen feet wide and eight inches thick; and 4 miles on part of the highway from Old Bridge to Mattawan, ten feet wide and nine inches thick.

1893~FIRST AMERICAN AUTOMOBILE

1893 FIRST AMERICAN AUTOMOBILE

TAYLOR ST.

SPRUCE AND FLORENCE STS.

What is generally accepted as the first American gasoline-engined automobile in the United States was given a short road test on the streets of Springfield, Massachusetts, by its builder, J. Frank Duryea, in September, 1893, as shown in the accompanying illustration. Actually gasoline engine automobiles were operated on the streets as early as 1889 by Henry Nadig, at Allentown, Pennsylvania, in 1890 by Gottfried Schloemer at Milwaukee, Wisconsin, and in 1891 by Charles H. Black at Indianapolis, Indiana. Construction on what was in reality a mechanized buggy was begun on April 4, 1892, by Charles E. Duryea and his brother, J. Frank Duryea, in the machine shop owned by John W. Russell & Sons, situated at 47 Taylor Street, Springfield. The phacton buggy had been purchased by Charles E. Duryea with $1,000 supplied by Erwin F. Markham, the financial backer of the project, who was to receive in return one-tenth of the future profits. Charles E. Duryea is credited with the original idea of a "horseless buggy" driven by a gasoline engine, the design of which had been suggested by C. E. Hawley, of Hartford, Connecticut. After having had made some drawings embodying the principal structural features of the car, Charles enlisted the aid of his brother Frank, then a tool maker.

Frank began at once to build the car according to the drawings submitted by his brother Charles, including the motor, transmission, and steering gear on the front axle. After repeated trials the original free-piston engine was found to be impractical. In September, 1892, Charles E. Duryea left Springfield for Peoria, Illinois, where he established a permanent residence and engaged in the bicycle business.

Following a siege of typhoid fever and further unsuccessful attempts to rebuild and run the original engine, J. Frank Duryea concluded reluctantly that the engine was a failure. About March 1, 1893, in agreement with his faithful sponsor, Erwin F. Markham, Frank began work on the drawings for a new engine based upon "sound and well-tried mechanical principles." This four-cycle, water-cooled motor, completed late in August, 1893, was installed in the reconstructed car together with Frank's design of an electric ignition, spray carburetor, governor, muffler, and hand cranking device; also, his framework, supporting the engine with transmission, pivoted on the front axle. Then, according to Frank, he and Mr. Markham "stood the vehicle up on end, resting on its rear wheels, to take it down the elevator, and it was left in the area between the Russell and Stacy buildings until after dark, when Mr. Markham's son-in-law, Mr. Bemis, brought a harnessed horse and we pulled it out to his barn on Spruce Street."

"A second item dated September 22, 1893, states definitely that 'first tests' have been made and shows clearly that the friction transmission had performed badly."

J. Frank Duryea began at once to redesign the transmission. After trying various arrangements of leather and rubber belts without success, he abandoned these materials as unsatisfactory and built an entirely new "gear and clutch transmission giving two speeds and reverse." This transmission having been installed, the "horseless buggy" was given another street trial which was described as a success in the January 19, 1894, issue of the Springfield Union.

Although it ran fairly well, the appearance of this car was so unsightly that it lacked commercial appeal. J. Frank Duryea, therefore, set to work and designed and built a vehicle more pleasing to the eye, which "had its first test on the top floor of the E.S. Stacy building late in 1894.

1893—THE OFFICE OF ROAD INQUIRY

One of the offsprings of the "Good Roads Movement" was the Office of Road Inquiry, United States Department of Agriculture, authorized by a statute enacted by the Fifty-second Congress, approved by President Benjamin Harrison on March 3, 1893. The statute read in part, "To enable the Secretary of Agriculture to make inquiries in regard to the systems of road management throughout the United States, to make investigations in regard to the best methods of road-making, and to enable him to assist the agricultural college and experiment stations in disseminating information on this subject, ten thousand dollars."

Secretary of Agriculture J. Sterling Morton instituted the Office of Road Inquiry on October 3, 1893, and appointed General Roy Stone, of New York, the former secretary for the National League of Good Roads, as Special Agent and Engineer for Road Inquiry in charge of the new organization, as shown in the accompanying illustration.

The office, in 1895, began the preparation of a Good Roads National Map showing all the macadamized and graveled roads in the country. Because the cost of hauling farm products averaged 22 to 31 cents per ton mile of road, traction tests were made at the National Road Parliament held at Atlanta, Georgia, on roads of different gradients and surface types. An increasing variety of studies were undertaken as better roads jumped land values from 30 to 800 percent. From August, 1898, to January, 1899, former Ohio State Highway Commissioner Martin Dodge, from Cleveland, filled the position of director vacated by General Roy Stone when on leave of absence while serving as a brigadier general on the staff of General Nelson A. Miles during the Spanish-American War. General Stone resumed his duties as Director of the Office on January 31, 1899, but resigned on October 23. Thereafter Martin Dodge was reappointed.

GENERAL ROY STONE

A testing laboratory, headed by Logan Waller Page, was installed by the office in 1900, in the basement of the Bureau of Chemistry building. The office cooperated with the National Good Roads Association and the Illinois Central Railroad, in 1901, in the first "Good Roads Train" which ran from New Orleans to Chicago as road experts gave lectures and built object-lesson roads along the route.

Pursuant to an Act of Congress, approved by President William McKinley on March 3, 1905, the Office of Public Roads Inquiries and the Division of Tests of the Bureau of Chemistry were consolidated on July 1 into the Office of Public Roads, with Logan Waller Page as director.

The Federal Aid Road Act passed by Congress was approved by President Woodrow Wilson on July 11, 1916. The Office of Public Roads and Rural Engineering on July 1, 1918, became the Bureau of Public Roads.

The Bureau of Public Roads on July 1, 1939, was transferred from the Department of Agriculture to the newly created Federal Works Agency and named the Public Roads Administration. On August 20, 1949, after being a unit of the new General Services Administration since July 1, the PRA was transferred to the Department of Commerce and designated once more the Bureau of Public Roads, now the Federal Highway Administration.

1893—FIRST BRICK RURAL ROAD

1893 FIRST BRICK RURAL ROAD

The first brick surface on a rural road in this country was laid on the Wooster Pike near Cleveland, in Cuyahoga County, Ohio. The four miles of brick pavement were built in the fall of 1893, on the old stage route between Cleveland and Wooster in Wayne County, now United States Route 42. The brick section was "reached by following out Pearl Street in the City of Cleveland, through the village of South Brooklyn to the second toll gate, four miles southwesterly from said village, where the four miles of road built in 1893" began.

According to civil engineer Jay F. Brown, who designed and superintended the construction of the improvement: "The road originally was 60 feet wide from fence to fence. We graded the central part of the road, a roadway 32 feet wide. On each side of the roadway was made a storm ditch of an average depth of 4 feet, 2 feet wide on the bottom, slopes, and banks, the slope being 1½ feet horizontal to 1 foot vertical. After the roadbed had been brought to a grade line and thoroughly finished, a line of drainpipe, 6-inch capacity, was laid along each side of the 32 feet; that is to say, a trench was dug 16 feet from the center line of the street to a depth of about 4½ feet below the grade line of the roadbed. The trench, after laying the pipe, was filled with stone, broken to 2½-inch size; ***** A drain of this kind was laid on each side of the roadbed, with outlets for water into every cross stream or ditch where it was possible to discharge the water, so frequent as not to overload the pipe in heavy storms. After the drains were put in, the strip of brick pavement above mentioned was laid close to one of the drains, leaving 24 feet width of dirt road for summer use. This dirt road was repeatedly rolled with a heavy roller until the upper foot or 2 feet of the crust of the roadbed became hard and solid."

The cross section of the improvement illustrated the "method of holding the brick in place alongside of the dirt road, instead of using a stone curbing. This plan, devised by me for this purpose, consisted of three courses of brick, standing edgewise, the first course flush with the top of the pavement, the second breaking joints and dropping two inches lower, the third 2 inches lower still, forming a stairwise bond for the brickwork in such a manner that a heavily loaded wagon can not catch and tear up the brick pavement. If a wheel runs off the pavement, it strikes the second course of curbing brick and runs along on that; but it is almost impossible for a wheel to cut through the broken stone filling which surrounds the curbing courses and protects them from the wear of heavily loaded wagons."

The brick surface on the Wooster Pike was one of the three road projects leading from the city of Cleveland selected by the commissioners for improvement. Built over a heavy white-clay soil, believed by many residents to present insuperable drainage obstacles, the brick road was completed successfully at a cost of about $16,000 a mile.

1896 ~ RURAL FREE DELIVERY

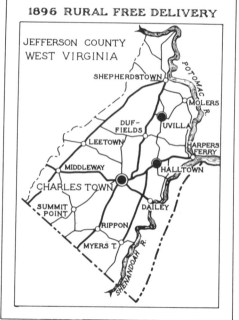

1896 RURAL FREE DELIVERY
JEFFERSON COUNTY WEST VIRGINIA

The first experimental routes for the rural free delivery of United States mail were established on October 1, 1896, during the administration of President Grover Cleveland, from Charles Town, the home of Postmaster General William L. Wilson, and simultaneously from Halltown and Uvilla, West Virginia. The routes were authorized first by an Act of Congress approved March 3, 1893, appropriating $10,000 for experimental rural delivery. "Surrounded at its birth by unfavorable auspices the path of rural free delivery was not a happy one." Delayed for three years by the contention of its opponents that the people should not be burdened "with such a great expense" the postal receipts in a few years after the service was inaugurated went far toward the payment of the cost. Furthermore, the demand for extension of the rural free delivery, shown in the accompanying illustration, gave a tremendous impetus to the Good Roads movement because a prerequisite for the service was a gravel or macadam road.

The Annual Report of the Post Office Department for the fiscal year ended June 30, 1899, contained the statement that, "There has been nothing in the history of the postal service of the United States so remarkable as the growth of the rural free delivery system. Within the past two years, largely by the aid of the people themselves, who, in appreciation of the helping hand which the Government extended to them, have met these advances half-way, it has implanted itself so firmly upon postal administration that it can no longer be considered in the light of an experiment, but has to be dealt with as an established agency of progress, awaiting only the action of Congress to determine how rapidly it shall be developed. *****

"That whenever the system has been judiciously inaugurated, with a sincere purpose to make it a success, it has been followed by these beneficial results:

I.—Increased postal receipts. More letters are written and received. More newspapers and magazines are subscribed for. So marked is this advancement that quite a number of rural routes already pay for themselves by the additional business they bring.

II.—Enhancement of the value of farm lands reached by rural free delivery. This increase of value has been estimated at as high as $5 per acre.

III.—A general improvement of the condition of the roads traversed by the rural carrier.

IV.—Better prices obtained for farm products, the producers being brought into daily touch with the state of the markets.

V.—To these material advantages may be added the educational benefits conferred by relieving the monotony of farm life through ready access to wholesome literature, and the keeping of all rural residents, the young people as well as their elders, fully informed.

"The recipients of the rural mail have to provide boxes and place them at convenient places along the line of road traversed by the rural carrier, so that he can deposit and collect the mails, if need be, without alighting from his buggy. Frequently, ***** similar to the accompanying illustration ***** seven or eight neighborhood boxes are grouped together like a lot of beehives at a crossroad corner, and the people living in houses perhaps half a mile or more back from the road watch for the daily passing of the carrier and come to the crossroad to collect or deposit their mails. *****

1897–FIRST OBJECT-LESSON ROAD

The first Federal object-lesson road was built by the Office of Road Inquiry of the United States Department of Agriculture at the entrance to the New Jersey Agricultural College and Experiment Station at New Brunswick. According to the Yearbook of the Department of Agriculture the work was done during the season of 1897 on "Nichol Avenue.—At New Brunswick, N.J., Nichol avenue leads from a main road or street in the southeastern section of the city to the State experimental farm. The portion improved is 660 feet in length, beginning at the street and ending at the gateway of the farm. The roadbed, with its general grade, was followed and dressed up with an American road machine. The shoulders were made up from the natural soil taken from the sides of the road, the soil being red clay, resting on red shale foundation. The stone used was the New Jersey trap rock, brought by rail from the Rockyhill quarries. The stone had to be unloaded from the cars and hauled to the crusher.

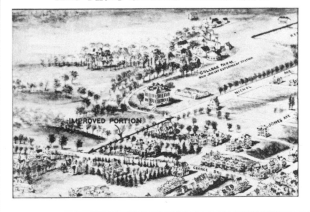

1897 FIRST OBJECT-LESSON ROAD

This extra hauling added to the cost. The crusher used was a No. 3 Champion. The distance from the crusher to the road where the stone was used was about 1-3/4 miles. The stone fell short and work had to be suspended for several weeks, so that much of the effect of filling or binding material was lost by not being rolled and wetted as it was put on. All the stone used was crushed, excepting about 20 cubic yards, which was purchased already crushed to finish the road. The total cost of the stone was $146; that of labor, sprinkling, rolling, and cartage $175; the cost per linear foot being 48-2/3 cents. The stone was laid 8 feet wide and 6 inches thick."

According to Director Roy Stone, of the Federal Office of Road Inquiry, the object-lesson road method was based upon the maxim that "Seeing is believing." Said Mr. Stone, "This method is especially valuable in teaching the importance of good roads and the possibility of obtaining them.

"In many parts of the United States the roads are torn up with the outcoming frost in the spring, soaked with autumn rains, frozen into ridges in winter, and buried in the dust in the summer, making four regular seasons of bad roads, besides innumerable brief 'spells.' For men accustomed all their lives to these conditions, it is hard to believe that country roads anywhere can actually be good all the year round. The lecturer on good roads, therefore, is listened to like one who tells fairy stories or travelers' tales of distant lands; but put down a piece of well-made macadam road as an illustration and let the people try it in all weathers and no lecturer is needed. The road speaks for itself, all doubts disappear, and the only question raised is how fast can it be extended and how soon can the improvement be general. *****

With respect to National object-lesson roads, Director Stone observed that, "It would greatly increase the value of the interstate roads and stimulate a general public interest in road building if some of these lines could be so connected or combined as to form in a measure, a national system." 79

1898 ~ THE DUST NUISANCE

Dust palliatives began to be applied to road surfaces as early as 1898. The first automobile drivers protected their eyes from dust particles by means of goggles, wore linen dusters to preserve their clothes or shielded themselves from dust clouds by riding in closed limousine cars. As the number of automobiles multiplied, attention shifted from the protective measures applied to the automobile and its occupants to the real source of the dust nuisance — the roadway itself.

James W. Abbott, Special Agent, Rocky Mountain and Pacific Coast Division, Office of Public Roads Inquiries, wrote, in the 1902 Yearbook of the United States Department of Agriculture, "Public attention was first called to the utility of crude petroleum oil in road betterment through experiments made by the county of Los Angeles in California in 1898, where 6 miles of road were oiled in that year under the direction of supervisors. The sole purpose of this work was to lay the dust, which, churned beneath the wheels of yearly increasing travel during the long dry season in that region, had become a most serious nuisance."

If proof is needed that oil as a road material antedated 1898 the evidence was presented by M. Meigs, U.S.C.E., of the U.S. Engineer Office at Keokuk, Iowa, in a letter to the editor of the Scientific American published on December 24, 1898, and quoted in part as follows: "As confirmatory of the value of oil on roads the following observations were made: [At a recent Good Roads Convention in St. Louis, Missouri] A gentleman from California said that near Santa Barbara, where he lives, they have oil wells and have used the oily sand from the borings to fill holes and ruts in the road, and in places the sand, has even been distributed over the roadway. In all these places the road is free from dust in the dry season (a great curse out there), and perfectly hard and firm in the wet season."

During the decade following 1898, the dust clouds from road surfaces had graduated from the nuisance class to the much more objectionable category of widespread destruction of water-bound macadam surfaces. Logan Walter Page, Director, Office of Public Roads, was quoted in the February 26, 1910, issue of the Engineering Record as stating, "It was quite generally believed by engineers before the introduction of motor traffic that at least 60 per cent of the wear on roads was due to the action of horses' feet; consequently, when the motor vehicle was first introduced, engineers and others having charge of roads welcomed this new type of vehicle, believing the soft pneumatic tire would have rather a beneficial effect.

"The first effect observed from fast motor traffic was an excessive amount of dust raised from the road surface, *****. The other injurious effect is the shearing stress of the driving wheels upon the road surface.***** The effect on a road subjected to much fast motor traffic is to denude the surface to a great extent of the fine binding material, which results in the larger stones in the road becoming loose."

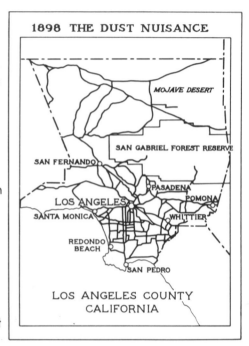

1898 THE DUST NUISANCE

MOJAVE DESERT

SAN GABRIEL FOREST RESERVE

SAN FERNANDO

PASADENA

POMONA

LOS ANGELES

SANTA MONICA

WHITTIER

REDONDO BEACH

SAN PEDRO

LOS ANGELES COUNTY
CALIFORNIA

1900
THE HORSELESS CARRIAGE

When the New Year's bells rang in the twentieth century the "horseless carriage" was still an experiment lacking even the dignity of an accepted name. It was some time before the word "automobile" found general favor. Many of the first "benzine buggies" were conversions from horse carriages. Their advent upon the roads was hailed with the derisive shout, "Get a horse!" from the throats of those lacking the vision to discern the signs of the times. These primitive mechanical contraptions were not only the butt of local jokes, and a source of feminine worry, but also required continual clumsy repair from beneath, illustrated herewith: Thus, when the first National Automobile Show was staged at Madison Square Garden, New York City, from November 3 to 10, 1900, in the same year that the Saturday Evening Post carried its maiden automobile advertisement, steam was considered the best power for motor vehicles.

The Automobile Club of America organized a road race from Springfield to Babylon and return on Long Island, New York, on April 14, 1900. The 50-mile event was won by a 5-horse power electric car driven by A.L. Riker. His time of 2 hours, 3½ minutes was 15 minutes faster than S.T. Davis, Jr., who piloted a steamer. A. Fisher in a gasoline car rumbled into third place 12 minutes later. The motor vehicle output, for the year 1900, totaled 1,681 steamers, 1,575 electrics, and 936 gasoline cars. The electrics were as fast as their rivals but were limited to a 25 to 30-mile run before recharging the batteries. The steamers could not start until the water boiled and they required expert mechanical handling. In a few years, therefore, preference shifted to the internal-combustion engine because fuel was inexpensive and the motor needed the least care and operating skill.

For the first time in the long history of the human race, horse power was destined to ignominy. The carriage builders were unable to stem the current of the new competition. Robert E. Olds with his Olds Motor Works introduced mass production by manufacturing 1,400 cars in 1900, 2,100 in 1902, and 4,000 in 1904. Their popularity was promoted by Gus Edwards' familiar song entitled, "My Merry Oldsmobile." The company adopted the slogan, "Nothing to watch but the road." These words presented an attractive appeal because it was not until a decade later that the automobile industry could produce a car that was reasonably certain of returning home over the insufferably rough roads.

Motor cars, in 1900, however, were restricted principally to wealthy people because of the prohibitive prices ranging from $3,000 to $12,000. In spite of the expense, motor vehicle registrations increased from 4, in 1895 to 14,800, in 1901, when New York State initiated a license fee which netted nearly a thousand dollars in revenue. The American craving for movement and speed has never been satisfied. T. H. Shevlin, in 1902, was arrested in Minneapolis, Minnesota, and fined ten dollars for driving a motor car faster than ten miles an hour.

1903—FIRST TRANSCONTINENTAL AUTOMOBILE TRIP

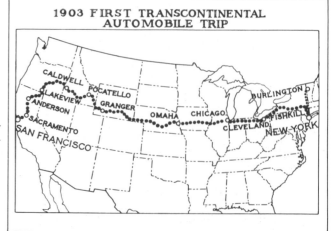

Two events, in the year 1903, were straws in the wind pointing to the trend of the automobile industry. One was the spectacular transcontinental trip made by Dr. H. Nelson Jackson, a prominent surgeon, from Burlington, Vermont, in a 20-horse-power Winton touring car which he purchased from a private owner in San Francisco by paying a premium above the $2,500 list price. Without fanfare, at his own expense, and motivated only by the ambition to be the first to complete the cross-country journey in a motor car, Dr. Jackson left the Golden Gate City on May 23, 1903, accompanied by his chauffeur Sewell K. Croker. The doctor avoided newspaper publicity because of a lingering doubt that the trip was possible. He arranged with a local agent to forward parts needed along the road. The car was loaded with supplies and equipment including a block and tackle. The route led from Sacramento across the Sierra Nevada Mountains, entered Oregon south of Lakeview, thence easterly across the Great Desert, the Continental Divide, the plains of Nebraska, the black mud of Iowa, through Illinois, Indiana and Ohio to New York City, traversing in all eleven States. A mishap in Wyoming is illustrated, herewith. From Nebraska east to the Hudson River "it rained every day and all day." The rope pulley changed defeat into victory. Mired eighteen times in one day in buffalo wallows they extricated themselves by using the rear axle as a windlass to power the rope tackle. Because of the detours made to avoid impassable trails and flooded areas, the 3,000 miles separating the Pacific and Atlantic coasts stretched to nearly 6,000 miles when the rugged doctor reached the New York metropolis, on July 26, 1903, after averaging more than 90 miles a day. On the road for 63 days his actual running time was 44 days. The tired traveler returned to his home in Vermont.

A second event, which at the time appeared insignificant, was the incorporation, on June 16, 1903, of the Ford Motor Company with a capital stock of $28,000. Nevertheless, this pioneer industrialist was destined to build an automobile to fit the pocket-book of the people. Manufacturing 1,695 cars in 1904, 1,599 in 1905 and 8,729 in 1906, Henry Ford in 1907, surprised the country with his announcement of standardized mass production. He proceeded to turn out 14,887 cars, during the same year, from his assembly lines. The small, low-priced "tin lizzie" became so popular that average automobile prices dropped from $3,000 in 1900, to $1,000 in 1911, and as low as $605 by 1916. The retail prices of the Ford 4-cylinder models, in 1917, varied from $345 to $645.

Meanwhile, horse-drawn buggy builders topped peak annual production of one million vehicles, in 1904, when the combined automobile makers crossed the 50,000 production mark. The total number of horses in the United States, however, did not begin to dwindle until after 1916 when the influence of the motor truck began to be felt.

1905~COAL TAR AND CRUDE OIL EXPERIMENTS

"During the spring and summer of 1905 the Office of Public Roads [United States Department of Agriculture] cooperated with Mr. Sam C. Lancaster, city engineer of Jackson, Tenn., and chief engineer of the Madison County Good Roads Commission, in making a series of careful experiments at Jackson, Tenn., to determine the value of coal tar for the improvement of macadam streets and roads. Tests were also made of the utility of crude Texas oil and several grades of its residue when applied to earth and macadam roads.

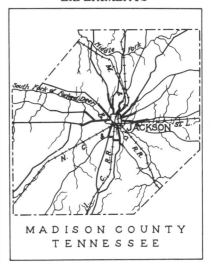

1905 COAL TAR AND CRUDE OIL EXPERIMENTS

MADISON COUNTY
TENNESSEE

"EXPERIMENTS WITH TAR

"The macadam streets in the business center of the city of Jackson were built originally of the hard siliceous rock known as novaculite. About the 1st of May, 1905, after 15 years of wear, repair of these streets became necessary. The old surface was first swept clean with a horse sweeper, so as to expose the solid pavement beneath. This was done because tar will not penetrate a road surface which is covered with dust or loose material. Next, the surface was loosened by means of spikes placed on the wheels of a 10-ton steam roller, the street reshaped, and new material added where needed. The road was then sprinkled, rolled, bonded, and finished to form a hard, compact, even surface, and allowed to dry thoroughly before either tar or oil was applied, for these substances can not penetrate a moist road surface. The best results are obtained when the work is done in hot, dry weather, and accordingly the tar was first applied in August. It may be well to add that the novaculite used in the construction of the roads is an almost nonabsorbent rock. *****

"After more than seven months, including the winter season of 1905~6, the tarred streets and roads are still in excellent condition. They are hard, smooth, and resemble asphalt, except that they show a more gritty surface. The tar forms a part of the surface proper and is in perfect bond with the macadam. Sections cut from the streets show that the tar has penetrated from 1 to 2 inches, and the fine black lines seen in the interstices between the individual stones show that the mechanical bond has been reinforced by the penetration of the tar."

"EXPERIMENTS WITH OIL

"Seven tank cars of oil, given by some Texas and Louisiana companies, were used at Jackson. It varied in quality from a light, crude oil to a heavy, viscous residue from the refineries. Over 7 miles of country road and several city streets were treated.

"At first, some of the lighter crude oils were applied with the same tank wagon that was used for the tar. Hose and brooms were used to spread the oil, and practically the same process was followed as with the tar. The oil soaked into the macadam very quickly and left no coating on top.

"More than seven months have now elapsed since the work was done. The light crude oil has produced little if any permanent results. *****

"The medium 'steamer oil' from Texas has given good results. ***** The road treated with the heaviest oil is entirely dustless. Teams passing from the bare macadam upon the oiled road show this, for the cloud of dust behind a wagon disappears at once, and the oiled roads can be cleaned."

1906–BITUMINOUS MACADAM EXPERIMENTS

The State Board of Public Roads of Rhode Island, in 1906, carried out experiments designed to find out the best method of construction for preserving road surfaces from disintegration caused by the wheels of swiftly-moving automobiles and for allaying the dust nuisance. The problem was described by James E. Owen in an address before the Good Roads – Automobile Convention in Atlantic City, New Jersey, in 1908, as follows: "Just consider what a road has to undergo. A heavy team comes tearing along with the horses' caulked feet, hammering and packing the stones for the heavy wheels to grind them to powder. Behind this comes a light buggy with a fast trotting horse and rubber tires, stirring up the loose material, then as a climax a six-ton motor car at 45 miles – excuse me, 21 miles – /New Jersey speed limit/ whizzes along throwing and hurling this loosened material into the gutter.

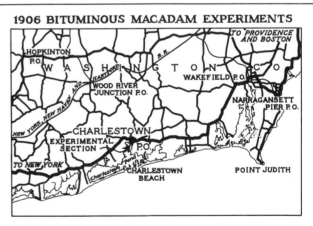

1906 BITUMINOUS MACADAM EXPERIMENTS

The Rhode Island experiments were described in the Transactions of the American Society of Civil Engineers, in December 1908, by Professor Arthur H. Blanchard of Brown University, "Among the methods used in the United States to alleviate the dust nuisance may be mentioned, sprinkling the surface with fresh water, salt water, a solution of calcium chloride, oils with a paraffin base, oils with an asphaltic base, oil of tar, oil emulsion, Westrumite, Dustoline, Asphaltoline, Tarracolio, and deliquescent salts. *****

"In the fall of 1906 a section of tar-macadam, 350 feet in length, was built in Charlestown, the location being on a curve of the interstate trunk line connecting New York, Narragansett Pier, Providence, and Boston. With the exception of the addition of the tar, the method of construction was similar to that used in building an ordinary macadam road. After the subgrade had been thoroughly rolled, the No. 1 broken stone, varying in size from $1\frac{1}{4}$ to $2\frac{1}{2}$ in. in longest dimension, was spread to a depth of 6 in. and rolled to 4 in. Tar, which had been heated to the boiling point in an ordinary tar kettle, was then sprinkled on the rolled surface by using dippers. The No. 2 stone, varying in size from $\frac{1}{2}$ in. to $1\frac{1}{4}$ in. in longest dimension, was next deposited on dumping boards and thoroughly mixed with hot tar by using rakes and shovels until every stone was completely coated. This mixture was applied on the No. 1 course to a depth of 3 in. and, after the tar had set, was rolled to 2 in. A thin coat of dust, which would pass through a $\frac{1}{2}$-in. mesh was then spread on the surface and forced by rolling into the No. 2 course to fill up the voids and provide a smooth surface. The quantity of tar used was 1.25 gal. per sq. yd. As this stretch of tar-macadam proved efficacious, it was deemed advisable, in 1907, to construct an experimental mile, this section being on the same interstate trunk line as the curve just mentioned. Although it will require from 5 to 10 years to determine the economical status and the efficiency of the tar-macadam road constructed by the method outlined, the perfect state of the surface of the 1906 and 1907 sections in the spring of 1908 influenced the State engineers to advocate the adoption of this method of construction."

1909~RURAL CONCRETE ROADS

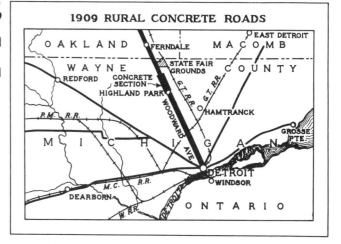

1909 RURAL CONCRETE ROADS

Credit for first surfacing a rural public road in the United States with Portland cement concrete pavement is conceded generally to Wayne County, Michigan. Construction work on the one-mile first unit, from 6 Mile to 7 Mile Road, on Woodward Avenue, leading from Detroit, Michigan, to the State Fair Grounds, was begun on April 20, 1909. The road was opened to the public on July 4, 1909, as shown in the accompanying illustration. "Automobilists, farmers and drivers of all sorts of vehicles" expressed their enthusiastic approval. "The total cost of the road, including administrative expenses, and share of depreciation on the plant and property of the commissioners, was $13,534.59."

According to a statement made by Edward N. Hines, Chairman of the Board of Wayne County Commissioners, "Woodward Avenue was selected as presenting the most trying conditions under which the road could be built. It is a continuation of Detroit's leading avenue, which, through the city and Highland Park village, is paved with asphalt, and is a thoroughfare to Palmer Park.

"***** The subgrade was tile-drained next to the car tracks. Labor was very favorable at the inception of the work, but with the advent of more prosperous times common laborers were hard to secure at $2 per day of ten hours, and teams were paid $5 per day."

Expansion joints were built into the pavement every 25 feet by placing in position a $\frac{1}{2}$-inch-thick strip of southern pine which was removed after the concrete had set. The space left was poured full of a heated filler composed of four parts of soft pitch and one part of refined Trinidad asphalt, plus 3 per cent of still wax. The concrete was cured under about a one-inch layer of sand or gravel kept damp for a period of seven days.

"Before preparing a specification for a concrete road, the commissioners visited nearby towns and inspected concrete crosswalks, bridge decks, also the concrete streets of Windsor, Ont."

This Canadian city in the Province of Ontario had been the first municipality in North America to pave streets extensively with Portland cement concrete, 32,000 square yards having been laid in 1907. The first concrete street pavement, however, in the Western Hemisphere, consisted of a 10-foot strip, 220 feet long, on the west side of Main Street, in the city of Bellefontaine, Ohio. This strip, which replaced the macadam paralleling the horse racks in front of the courthouse, was begun on June 13, 1891. A small stretch of concrete alley previously had been laid in 1890, in the city of Connersville, Indiana. The earliest Portland cement concrete pavement is attributed to Inverness, Scotland, in 1865.

The first pavement on a country road of Portland cement concrete was opened to traffic on October 10, 1908, on private right-of-way owned by the Long Island Motor Parkway Company, as a race course for the William K. Vanderbilt, Jr., automobile cup race. The 11-mile section, with superelevated curves, was paved with steel-wire-mesh Hassam-type reinforced concrete 24 feet wide, in two courses, with a total thickness of five inches.

1911–THE MOTOR PATHFINDERS

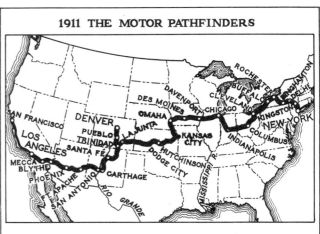

1911 THE MOTOR PATHFINDERS

The first transcontinental motor-truck trip made, in 1911, demonstrated the usefulness of the motor vehicle as a freight carrier. The motor truck now was ready to put on long pants and do a man's job. The journey was performed in a four-cylinder, 37 horse-power, gasoline-engine powered truck, called the "Pioneer Freighter," built by the Sauer Motor Car Company of New York. The vehicle was equipped with a chain drive from the counter shaft to the rear axle. The transmission allowed four rates of speed. The front wheels were shod with single 36-inch diameter and and the rear wheels with dual 42-inch diameter solid rubber tires. The 24-gallon gasoline tank permitted a touring range of 168 miles over good roads where seven miles could be covered with a gallon of fuel. The weight of the unloaded body and chassis was about three tons. The load, consisting of lumber for bridging creeks and soft roads plus camp equipment and supplies, weighed another three and one half tons. The loaded weight, therefore, totaled 14,000 pounds. The truck was piloted by A. L. Westgard, representing the Touring Club of America.

The westward leg of the tour began on March 4, 1911, when the truck rolled from Denver, Colorado, thence running southwest, shown in the illustration, through Santa Fé, New Mexico, and Phoenix, Arizona, to Los Angeles, California, entered on May 9, after traveling 1,450 miles in 66 days of which 53 days were consumed in travel. The running time of 444 hours and 20 minutes averaged a speed of 3.26 miles an hour–about as fast as a man could walk. A block and tackle was used to haul the truck through desert "soft spots" and to "submarine", or ford, the streams. From Los Angeles the machine was driven to San Francisco and shipped by railroad back to Pueblo, Colorado, south of the Denver starting point.

The eastward reversed leg of the journey began on June 12, 1911, when the truck left Pueblo under its own power, arriving in Kansas City, Missouri, on June 18, after traveling 700 miles in 70 hours actual running time, an average of 10 miles an hour, including 8-hours delay when the heavily-laden truck crashed through a high wooden-trestle bridge west of Hutchinson, Kansas. The journey ended at New York City. This endurance run as well as the first Motor Truck Show convened at Madison Square Garden, New York City, in 1911, served notice that the freight motor carrier had grown in stature to economic importance.

After World War I broke out in Europe on August 4, 1914, motor truck and tractor manufacture expanded to speed transportation of war materiel to home ports for shipment overseas and to increase farm productivity. The total horse and mule population of the United States reached a peak of 26,723,000, in 1918, the year of the Armistice. Thereafter, animal breeding joined the retreat of wagon and carriage production.

The atrocious condition of the cross-country trails revealed by the Sauer pathfinder motor truck and by the transcontinental automobile tourists who preceded it, revealed the need for a surfaced east-west highway.

1913
FIRST POST-ROAD PROJECT

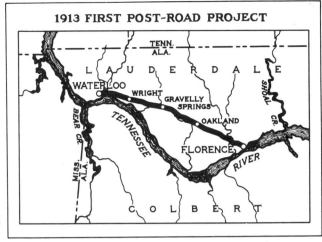

1913 FIRST POST-ROAD PROJECT

The first post-road project built from funds provided by the Post Office Appropriation Bill (C. 284, 37 Stat., 269, 299), approved August 24, 1912, for the fiscal year 1913, consisted in the improvement of the Waterloo Post Road leading from Florence, Lauderdale County, Alabama. The bill appropriated the sum of $500,000 to be expended by the Secretary of Agriculture in cooperation with the Postmaster General for improving the condition of the roads in the several States over which rural free delivery of the mails "is or may hereafter be established." The act provided that the Federal Government should contribute one-third and the local or State government two-thirds of the cost of the projects. Under this plan each of the forty-eight States was to be allotted $10,000 and the remaining $20,000 was to be used for administrative expenses and contingencies. It was expected that the average expenditure would approximate $600 a mile so that each State would complete 50 miles.

According to the Joint Report on the Progress of Post Road Improvement to the 63d Congress, 1st Session, known as House Document No. 204, dated August 23, 1913, active work began on this Alabama post road, approximately 30 miles long, on May 26, 1913. The exact length of the preliminary survey was 29.96 miles. By August 23, 1913, the final location survey had been completed over 14.56 miles of this dirt road project, the previous bad condition of which is shown in the accompanying illustration. A $30,000 joint fund had been set aside for this unit and a dirt road in the State of Iowa.

The Annual Report of the Director of the Office of Public Roads, of the United States Department of Agriculture, who had jurisdiction over the improvement, stated that during the fiscal year ending June 30, 1914, "Only one project, that in Lauderdale County, Ala., was completed *****. The work on this project consisted of grading 72,240 square yards of road at a cost of $25,781.09, and surfacing a part of the road at a cost of $2,166.05."

From the foregoing report it is apparent that the project, originally intended to be only a dirt road, was later surfaced over a part of its length. Other records give the width of this gravel surfacing as 12 feet laid on the graded earth road including the installation of culverts and small bridges.

The improvement of the post roads, pursuant to the Post Office Appropriation Bill, approved August 24, 1912, provided the necessary training for the engineers of the Office of Public Roads. Later this experience was put to good use in the administration of the Federal Aid Road Act.

The reader should not conclude from the foregoing statements that the Waterloo Post Road in Lauderdale County, Alabama, was the first post road improved in this country with Federal funds, for such is not the case. As a matter of fact, nearly a century before, on May 20, 1826, Congress appropriated $6,000 "For the repair of the post-road in the Indian country, between the Chattahoochee (River) and Line Creek, in the State of Alabama, to be expended under the direction of the Postmaster General."

1913~THE LINCOLN HIGHWAY

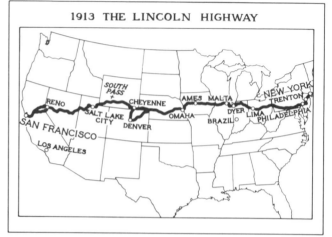

1913 THE LINCOLN HIGHWAY

The Lincoln Highway was conceived with the aim of building a great transcontinental object-lesson road at a time when there were no improved long-distance highways in the United States. In 1913 no State, nor probably any county, could boast of a completed highway system.

As the motor vehicles rolled from the assembly lines and out into towns and through the countryside in growing numbers, Carl G. Fisher, who had been associated with the automobile industry from its infancy, visioned the possibilities of a Coast-to-Coast Rock Highway. On September 6, 1912, Mr. Fisher presented the project to a group of automobile manufacturers and received their enthusiastic approval. Within a short time 4 million dollars were pledged toward the construction of the road. Then followed the customary discussions preceding the first formal meeting of the organizers held on July 1, 1913, at the national headquarters in Room 2115, Dime Savings Bank Building, Detroit, Michigan. There the Lincoln Highway Association was chosen as the name of the organization.

Henry B. Joy, the president of the Packard Motor Car Company, was elected as the first president of the fledgling Lincoln Highway Association. Mr. Joy and his Board of Directors decided that it was the paramount duty of the association to build on the shortest, best and most direct route across the midsection of the country from New York to San Francisco, illustrated on the map (right). The South Pass crossing of the Rocky Mountains was agreed upon as the most practicable route over the Continental Divide. The original projected length of 3,389 miles later was shortened.

To arouse countrywide interest in the Lincoln Highway, an automobile caravan of 70 persons left Brazil, Indiana, on July 1, 1913. The tour reached Los Angeles without serious mishap 34 days later. From its inception the backers of the Lincoln Highway Association had in mind the promulgation of a broad educational program emphasizing the need for better roads. With this objective in view, a program of "seedling miles" was inaugurated with the assurance that each unit of good road would bear fruit in many more miles of a similar pattern.

The passage of the Federal-aid Road Act on July 11, 1916, relieved the Lincoln Highway Association of the burden of advocating for better roads. With the general public conscious of this need, the activities of the Association shifted to spelling out proper road design and methods of construction. To accomplish this purpose the blueprints for an Ideal Section of the Lincoln Highway were adopted after consultation with leading highway engineers. The pavement was to be of cement concrete, 40 feet wide, capable of carrying four lanes of traffic and flanked on each side by pedestrian side paths, all included within a right of way 110 feet wide. The roadway was designed to carry automobiles at an average speed of 35 miles an hour and motor trucks at 10 miles. Work began on July 7, 1922, on the 1-1/3-mile Ideal Section situated in Lake County, Indiana, beginning at Dyer on the Illinois-Indiana boundary, 33 miles south of Chicago. The concrete paving, shown in the accompanying illustration, was completed in December.

With the establishment of the numbered system of United States highways, in 1925, the Lincoln Highway lost significance in common with other named roads. For the greater portion of its length, the Lincoln Highway coincided with the present United States Route 30.

1915
MITCHELL'S POINT TUNNEL

Mitchell's Point, or Storm Cliff tunnel, on the famous Columbia River Highway in the State of Oregon, probably was the first important highway tunnel built in this country. Completed more than four thousand years later than the first tunnel of historic record used by the King and priests of Babylon to walk back and forth under the Euphrates River between the royal palace and the temple, the Mitchell Point tunnel added another arrow to the quiver of the highway engineer. Outranked in length by tunnels built by the Romans, the greatest ancient exponents of this type of construction, the Mitchell's Point tunnel embraced picturesque features rivalling the modern Axenstrasse bordering Lake Lucerne, in Switzerland.

Tunneling began in this country as early as 1821, on the Schuylkill River navigation canal above Auburn, Pennsylvania, and increased in scope with the extension of the canals and later steam-railroad systems. The first railroad tunnel was opened

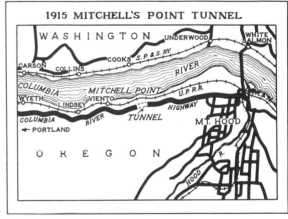

at Staples Bend, Pennsylvania, on the Allegheny Portage Railroad, in 1831-33. Subsequently tunnels became quite common on railroad and city subway development where the object was to provide the greatest transportation service at the least cost.

To the utilitarian features of tunnels, Samuel Christopher Lancaster, the consulting highway engineer in charge of the design and construction of the Columbia River Highway, added emphasis upon the social values of the picturesque. Mr. Lancaster visioned a magnificent scenic highway.

John A. Elliott, who had been assigned to the location of the Columbia River Highway in Hood River County by Oregon State Highway Engineer Henry L. Bowlby, called the attention of Mr. Lancaster to the possibilities for the tunnel with the remark, "I believe I have found a place to duplicate the Axenstrasse in Switzerland." Mr. Lancaster replied, "Do it if at all possible. It will be one of the finest things on the highway."

Mitchell's Point was a rugged ridge of basaltic rock projecting from the south bank of the Columbia River about five miles downstream from Hood River, Oregon. The upper portion of the ridge, 1,100 feet in elevation, was separated from the 400-foot-high lower ridge by a low saddle 250 feet in height. The base of the rocky point beside the river was skirted by the main-line tracks of the Oregon-Washington Railroad and Navigation Company. A pioneer, narrow, precipitous wagon road, with grades as steep as 23 per cent, built in the 1860's traversed the saddle. A preliminary reconnaissance survey proved that a tunnel through the cliff would be a mile shorter and much less costly to build than a road across the saddle. After survey crews, suspended from ropes fastened two hundred feet above, had fixed the location of the tunnel, actual construction work was begun on March 31, 1915, and completed on November 10 of the same year. The tunnel was eighteen feet wide for a height of ten feet above the surface of the roadway. Above this rectangular bore the cross-section converged into a semicircular dome with a radius of nine feet. Throughout the 390 feet of tunnel, alined near the middle on a 10-degree curve with a central angle of thirteen degrees, there were spaced symmetrically five windows each sixteen feet wide. The entire tunnel was blasted with dynamite from the solid rock with the utmost care in order not to disturb the supporting rock framing the windows. The only artificial elements were the masonry railings across the windows, ninety feet above the railroad tracks. A reinforced concrete viaduct, across a talus rock slide, 208 feet in length, connected the west portal of the tunnel with solid ground on the adjoining rock ridge.

1916~ THE STATE LINE

State boundaries, in 1916, often seemed like canal locks separating a high level of highway improvement on one side from a lower rate of progress upon the other. The situation, depicted in the illustration, was not unusual. A traveler driving over a smooth, hard-surfaced, well-kept State highway might arrive at a State boundary only to have his journey halted abruptly by a muddy, pot-holed earth road in the adjoining common-wealth. A diagnosis of the trouble showed the need for coordinating the main interstate roads. A similar condition had obtained, in 1891, when New Jersey enacted the first State-aid road law. At that time the bicyclists, organized into the League of American Wheelmen, were attempting to extend their radius of travel from the towns across the county boundaries which hitherto had marked the extent of a day's travel for a horse and wagon to and from the county seats. During the following 15 years State after State passed local-aid laws and county-road networks were expanded to the State borders only perhaps to come to a dead end. The 3,617,917 motor vehicles in the country, in 1916, were seeking to extend the radius of travel far beyond the previous limits of the bicycle. The pressure was nationwide because nearly 61 per cent of all the automobiles manufactured at that date were Fords. The low price of these cars gave them a wide range of usefulness.

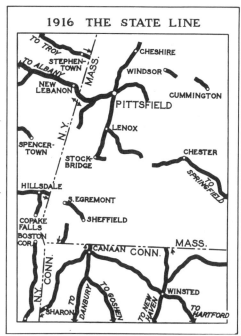

The Federal Aid Road Act, signed by President Woodrow Wilson, on July 11, 1916, was designed to coordinate the main interstate roads in the same manner that the previous State-aid road laws had integrated the county road systems. This law was the first comprehensive act of our Government aimed at the establishment of a nationwide system of interstate highways. The Act was based upon Article 1, Section 8, Clause 7 of the Constitution of the United States which provides that, "The Congress shall have the power * * * * * To establish Post Offices and Post Roads." The law recognized the necessity for the consent of the States based upon their sovereignty. The issue of States' rights was at a high pitch, in 1822, when President James Monroe vetoed the "gate bill" providing for the collection of tolls for the repair of the National Pike.

After the lapse of nearly another century the authority of the central Government had become established more firmly and the matter of States rights was viewed in clearer perspective. Thus, the time was ripe for the passage, in 1916, of the Federal Aid Road Act which provided for the construction of "rural public roads" and defined them as "any public road over which the United States mails now are or may hereafter be transported." The Act further stated that Federal contributions "shall not exceed fifty per centum of the total estimated cost" of each project and that each State shall "maintain the roads constructed under the provisions of this Act." This last provision forestalled revival of the disagreement which permitted the National Pike to fall into a condition of disrepair because it was held that repair work would necessitate Federal jurisdiction over the roadways and be an invasion of States' authority.

1918
FIRST FEDERAL-AID ROAD

The first unit completed under the authorization of the Federal Aid Road Act, approved by President Woodrow Wilson on July 11, 1916, was California Federal-Aid Road Project No. 3, situated in Contra Costa County and known locally as the "Alameda County boundary to Richmond road." The project, shown in the accompanying illustration, was 2.55 miles in length and extended from Albany (Alameda County boundary) to Richmond in Contra Costa County.

The work consisted of grading the roadbed, draining and installing culverts flanked with concrete headwalls, and laying a Portland cement concrete base, in the proportions of 1:3:6, with a width of 20 feet and a thickness of 5 inches, surfaced with a bituminous concrete top (Topeka mix) 1½ inches in thickness. Construction began officially on September 1, 1916, and the certificate of completion was issued by the District Engineer of the Bureau of Public Roads of the United States Department of Agriculture, on January 30, 1918. The total cost of the project, including the money allotted by the State, was $53,938.85.

The Federal Aid Road Act, based upon the clause in the Constitution of the United States which empowered Congress to establish post roads, stated that a "'rural post road' shall be construed to mean any public road over which the United States mails now are or may hereafter be transported, excluding every street and road in a place having a population, as shown by the latest available Federal census, of two thousand five hundred or more."

The act was intended by its sponsors to accomplish on a National scale what the State-aid road laws, initiated by the State of New Jersey, in 1891, had authorized at the county level. That is, the act was worded to promote the improvement of a nationwide system of free roads under the direction of experienced highway engineers. To insure this purpose the act provided, "That the Secretary of Agriculture is authorized to cooperate with the States, through their respective State highway departments, in the construction of rural post roads; ***** Provided, That all roads constructed under the provisions of this act shall be free from tolls of all kinds. *****."

Having established as a prerequisite for the receipt of Federal-aid funds that each State should have a highway department, staffed with a corps of trained highway engineers, the act proceeded to equip the Secretary of Agriculture with a similar engineering organization.

To spread the mileage of roads improved over the greatest possible area of the United States the act established a maximum Federal-aid contribution of, "$10,000 per mile, exclusive of the cost of bridges of more than twenty feet clear span." To insure the most equitable distribution of the appropriations among the States, the act provided for the division of the funds, after the deduction of the administrative funds as follows; one-third in ratio of the area of the State to the total area of the United States; one-third likewise in respect to the population of the State; and similarly one-third with respect to the mileage of rural delivery and star routes in each State.

1918 FIRST FEDERAL-AID ROAD

1920~BEGINNINGS OF HIGHWAY RESEARCH

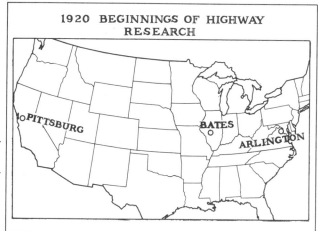

1920 BEGINNINGS OF HIGHWAY RESEARCH

A nationwide program of highway research of large-scale proportions was launched in 1920. By October, 3,191 miles of Federal-aid projects had been completed at an aggregate cost of $46,740,472 of which the Federal funds amounted to $20,900,014. The expanding Federal-aid program, however, was not keeping pace with the requirements of the 9,231,941 motor vehicles registered by that year. These cars were rolling from the manufacturers' assembly lines out upon the roads in a steady stream. Their rubber-tired wheels were destroying the light water-bound macadam surfaces built for horse-drawn carriages and wagons with steel-tired wheels. The pneumatic-tired automobiles loosened the broken stone and the fine binding material was scattered by the wind. The impact of the heavy solid-rubber-tired trucks pounded the thin surfaces to pieces and shook and often broke through the flimsy wooden bridges and the slender-trussed steel structures. Clearly something had to be done and at once.

Test information was needed with respect to the supporting power of the various subgrade soils, the direction and value of the stresses induced in rigid and flexible pavements by the impact of motor vehicles, the effects of expansion and contraction of road surfaces caused by variations in temperature, the wear of traffic upon pavement surfaces, the distribution of loads upon bridges and many other factors.

Technical periodicals, in 1920, were replete with articles by prominent highway engineers stressing the dearth of research data. Appreciating the urgency of the matter the American Association of State Highway Officials organized a test and investigation committee. After scanning the highway research horizon this group reached the conclusion that isolated studies by State and Federal road building authorities, universities and other agencies, while highly commendable, would fail of accomplishing the desired objectives unless coordinated into a nationwide program of highway research. Not only should information be exchanged with respect to results obtained and studies in progress but an agreement should be reached as to the nature of the problems pressing for solution. Thus, a conference was attended by representatives of the State highway departments, the Bureau of Public Roads, United States Department of Agriculture, with the engineering division of the National Research Council acting as the coordinating agency. Out of these talks came the Highway Research Board of the National Research Council, organized to provide a clearing house and a forum.

Also, in 1920, the Bureau of Public Roads speeded its field tests on the large circular track containing different sections of pavement at Arlington, Virginia. Imposing experimental roads were constructed by the Illinois State Highway Department at Bates, southwest of Springfield and in 1921, by the Columbia Steel Company cooperating with the California Highway Commission at Pittsburg east of San Francisco. The Bureau of Public Roads intensified its laboratory tests and initiated a country wide field study of subgrade soils. Universities became beehives of research activity.

The branches of this research tree soon grew to healthy proportions.

1920~FIELD SUBGRADE SOIL STUDIES

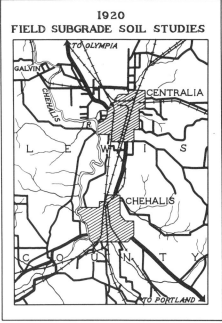

The "field inspections of road failures produced by poor subgrade conditions" were initiated by a memorandum to the District Engineers, dated April 20, 1920, signed by Thos. H. Mac Donald, Chief of the Bureau of Public Roads, United States Department of Agriculture. The District Engineers were informed that, "The Division of Tests at Washington has begun an investigation in the vicinity of Washington with this idea in view but little can be accomplished along this line unless a nationwide research be made obtaining the services of various cooperating agencies. ***** It is obvious that if we are to advance our ideas regarding road construction we must begin with the road foundation and must gain accurate information as to what properties of soils make them good or poor soils for foundation purposes.

"***** I would suggest that you designate someone in your district to make an inspection of a number of road failures due to poor subgrades, whether due to poor drainage or to peculiar soil conditions, working in cooperation with the State highway departments and also, wherever possible, with the State geologists."

This memorandum had its genesis in the widespread damage which the heavy motor-truck traffic during World War I inflicted upon surfaces built during preceding generations to sustain horse-drawn vehicles. The editor of Public Roads commented in June 1918, "Apparently the point has been reached where the demands of traffic have exceeded the strength of the average road to meet them. ***** From horse-drawn vehicles with a concentrated load of probably 3 tons at most, traveling at the rate of 4 miles an hour, sprung almost overnight the heavy motor truck with a concentrated load of from 8 to 12 tons, thundering along at a speed of 20 miles an hour. The result ? The worn and broken threads that bind our communities together. The solution ?"

Pressed for answers to the problem, leading State highway officials suggested maximum limits for the total weight of motor trucks, a $500 fee for the heaviest trucks, strengthening the existing surfaces, improving drainage conditions, and better maintenance. These were immediate stop-gap measures aimed at keeping a continuous flow of war traffic in operation between the munition, ordnance and supply plants and the seaports. Then the war ended suddenly with the Armistice of November 11, 1918, and there followed at once a public clamor for "improved roads at once and everywhere."

It was to implement this program that the Chief of the Bureau of Public Roads issued the directive for the field study of highway subgrades on April 20, 1920. As a result of this order District Engineer C.H. Purcell appointed A.C. Rose to make a study of subgrade conditions in the States of Oregon and Washington. From these investigations there eventuated the discovery that good subgrade soils were those having a field-moisture-equivalent value of less than 20 per cent, doubtful soils ranged from 20 to 30 per cent and bad soils exceeded 30 per cent. Rapid field methods for identifying the quality of subgrade soils were devised which were printed in the August 1924, July and September 1925 issues of Public Roads.

1920
THE CONSOLIDATED RURAL SCHOOL

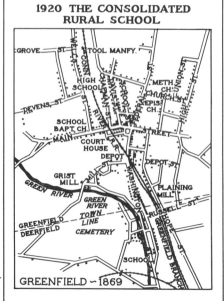

1920 THE CONSOLIDATED RURAL SCHOOL

GREENFIELD – 1869

The total number of one-teacher schools in the United States began to decline sometime in the decade between 1910 and 1920. The eclipse of the "little red school house on the hill" by the consolidated graded school, shown in the accompanying illustration, staffed with a corps of teachers familiar with the needs of the various age groups, and centrally located in a rural region, was speeded by the programs of Federal-aid and State highway construction which began to gain momentum soon after the close of World War I.

The one-teacher rural schools had their origin early in Colonial America as the expansion of the frontier to the west made it necessary to establish outlying centers of instruction beyond convenient walking distance to the town schools. These country schools had been growing in number long before the first legal basis was established, in 1789, by the General Court of Massachusetts, as follows: "And whereas, by means of the dispersed situation of the inhabitants of the several towns and districts in this Commonwealth, the children and youth cannot be collected in any one place for their instruction, it has thus become expedient that the towns and districts, in circumstances aforesaid, should be divided into separate districts for purposes aforesaid, be it enacted ***** "

Thus our forefathers decided that if the children could not walk to the schools then the schools must be brought within walking distance of the children. That many of these one-teacher schools maintained by poor and sparsely-settled school districts were inefficient was attested as early as 1844 by Superintendent Dix of the State of New York who said, "In feeble districts cheap instruction, poor and ill-furnished schoolhouses, and a general languor of the cause of education are almost certain to be found. *****

To overcome the deficiencies of the "little red schoolhouse," consolidation of one kind or another has been in progress since the beginning of public instruction in this country. It was not until 1869, however, that the first consolidation occurred which involved the reorganization of school districts and the transportation of pupils. In this instance Greenfield, Massachusetts, abandoned three "district" schools and transported the children at the public expense to a new centrally-located brick building. After the town of Montague, Massachusetts, followed suit a few years later with a consolidation of their own, the innovation spread throughout the country. By 1909, laws in half of the States encouraged the union of school districts, and 18 States had enacted legislation permitting the expenditure of school funds for the transportation of the pupils. In the ensuing years consolidated central schools developed rapidly, growing in number from about 5,000 in 1916, to 17,531 in 1936. During the concurrent 18-year period the number of one-teacher schools decreased about one-third, dropping from 196,037 in 1918, to 131,101 in 1936. According to these figures, 8 or 9 one-teacher schools disappeared from the country educational landscape every day.

The number of children brought to school in publicly-owned or operated vehicles multiplied six fold in the two decades between 1916 and 1936.

1921
PNEUMATIC TIRE IMPACT TESTS

Observers traced the widespread destruction wrought upon the highways of this country, in 1918, to the impact of the solid-rubber-tired wheels of motor trucks which pounded over the roadways in an unending stream flowing from the interior to the seaports. The light surfaces designed for horse-drawn traffic were broken into thousands of pieces, pulverized and scattered to the four winds of the heavens. Deep ruts, continuous over long distances, were cut through the bituminous macadam surfaces on such a main route as the Washington-Baltimore Boulevard. Light concrete pavements elsewhere were smashed to smithereens. Thus, when World War I ended, the renewed Federal-aid highway program had for its first objective the reconstruction of the main highways of the country, demolished since we began to ship supplies to our later allies overseas, early in 1914.

Tests were begun at the Experimental Farm of the Bureau of Public Roads at Arlington, Virginia, to measure qualitatively these impact forces delivered by the wheels of motor trucks. A preliminary report on these original tests was published in the July, 1919 issue of Public Roads. A 3-ton United States Army Class-B truck, equipped with solid rubber tires was employed in these initial experiments. The magnitude of the impact was measured by the deformation of a copper cylinder one-half inch long. The cylinder was placed in a jack upon which was dropped the wheel of the motor truck loaded with various tonnages and traveling at different rates of speed.

These early impact studies revealed the elusive nature of the impact force which necessitated the construction of more refined measuring instruments such as the accelerometer. In 1921, A.T. Goldbeck, Engineer of Tests for the Bureau of Public Roads, wrote, "When we consider the design of any structure built to carry loads we must know with reasonable accuracy what will be the maximum load. The motor truck as it is commonly built at the present time has its weight distributed on four wheels, with as much as 80 per cent of the gross load carried on the two rear wheels. The road surface, then, must support four concentrated loads, the maximum concentration being not uncommonly 12,000 pounds on a single wheel."

Later, as the tests indicated more clearly the relationship between concentrated load and pavement thickness, the 9,000 pound maximum wheel load was adopted as the economic standard upon which to base the design of a nationwide system of highways. Clifford Older, Chief Highway Engineer of the State of Illinois, had demonstrated that the corner was the weakest portion of a concrete pavement slab. He had devised a formula which evaluated a 750-pound-per-square-inch fiber stress when a 9,000-pound wheel load was applied to a concrete slab six inches in thickness. Mr. Older had developed his formula during the extensive field tests of various types of pavement on a 2-mile test road at Bates, Illinois.

1921—THE APPALACHIAN TRAIL

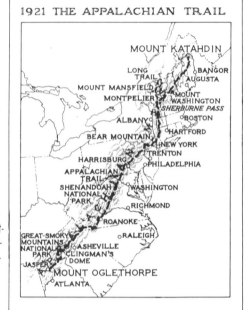

The origin of the Appalachian Trail may be traced to the aspirations of enthusiastic hikers who longed for a long-distance route through the wilderness over which they could expend their surplus energies and perhaps recapture some of the treasured memories of our pioneer past. The early Indian trails and such famous routes as the Oregon Trail, the Santa Fé Trail and the Natchez Trace were topics of fireside conversation delighted in by campers. From 1800 to 1850, when 85 per cent or more of our population lived in the country, many a sickly city dweller had regained health because of the bracing air and the vigorous exercise incident to a journey over one of the pioneer roads. Perhaps the recollections of those halcyon days kindled the imagination of the nature lovers of 1921 when more than half of the citizens of the United States were domiciled in urban areas housing 2,500 or more people.

The seed of the Appalachian Trail idea was planted by Benton MacKaye, a forester, philosopher and dreamer living in Shirley, Massachusetts. His plan, entitled, "The Appalachian Trail, an Experiment in Regional Planning," was published in the October, 1921, issue of the Journel of the American Institute of Architects. Interest in the project soon spread from New England to Pennsylvania. The initial enthusiasm waned, however, before much work was done to transform the project from the status of the printed page into reality. The fallen torch, in 1926, was picked up by Arthur Perkins, a retired lawyer of Hartford, Connecticut. Before the driving force of his dynamic personality the Appalachian Trail was translated from a fireside dream into disconnected footpaths through the mountain wilderness which were joined into one continuous 2,050-mile-long trail by 1937.

The northern extremity of the Appalachian Trail is anchored upon Mount Katahdin (elevation 5,267 feet), an immense granite monolith looming above the wilderness of central Maine.

The Appalachian Trail was built and is being maintained by the 26 hiking clubs in the Appalachian Trail Conference in cooperation with State and National Forest and Park officials. Private landowners also help in the work of clearing and marking the trail. On October 15, 1938, Federal officials approved the new type of recreational area known as the Appalachian Trailway consisting of a zone one mile wide on each side of the Appalachian Trail across National Forests and Parks.

The first walker to cover the continuous length of the Appalachian Trail was Earl V. Shaffer of York, Pennsylvania. He left Mount Oglethorpe, Georgia, on April 14, 1948, and after spending 123 nights along the route and averaging daily 17 miles of hiking, he reached Mount Katahdin on August 5. The Appalachian Trail has advanced the opportunities for outdoor recreation throughout its entire length ranging from the vigorous skiing and other outdoor sports in the White Mountains of New Hampshire to the restful relaxation in the comfortable summer homes in the rugged but picturesque altitudes of North Carolina.

1922~SNOW REMOVAL

The removal of snow from the main highways of the country with motor-driven equipment began in earnest during the winter of 1922–23 when 27,096 miles of snow-covered roads were opened for travel by the State highway departments. Previous to this season there had been units of motor-driven equipment scattered here and there, but the bulk of the work was performed by means of primitive wooden plows drawn by horses or with hand shovels. Systematic snow-removal operations on a nationwide scale did not gather momentum until after World War I, when the rapid increase in motor vehicle registrations during the early 1920's brought about a widespread agitation for principal highways open for travel the year round. The mileage of the highway systems cleared in the 36 heavy-snowfall States—those with an annual depth of 20 inches or more, as shown in the map above – increased by about one-half for each year up to and including 1925–26. Thereafter the mileage increase dropped to 15 per cent, indicating that the State highway departments had serviced nearly all the principal routes of travel and that further extensions in snow-removal mileage would be accounted for by the normal growth in traffic elevating secondary roads to the status of primary thoroughfares.

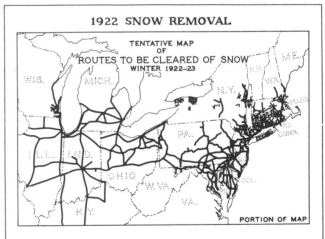

The motor-driven equipment in these early days consisted in general of four types: (1) Straight-blade plows attached to motor trucks; (2) V-shaped plows mounted upon motor trucks; (3) V-shaped plows fixed to motor-driven tractors and (4) motor-driven rotary plows of the transverse or lateral-fan type. The first class specified above were economical for use in areas of light snowfall not exceeding 6 inches in depth on the level or 15 to 18 inches in short drifts. The plows of the second class were adapted to a 15-inch mantle of light snow on the level or 18- to 30-inch drifts of moderate length. The equipment of the third class provided a heavier tool capable of removing wet snow compacted to a 12-inch to 3-foot covering on the level or to buck aside light drifts as deep as 4 feet or more. The rotary machines of the fourth class were restricted to areas of excessive snowfall varying from 2 to 9 feet, such as mountain passes, or to opening paths through long, deep drifts which resisted lateral movement by a V-plow.

The cost of snow removal over a period of years was found to be much more expensive than comparatively simple preventive measures. Another example of an ounce of prevention being worth a pound of cure. Drifts could be avoided in deep cuts by erecting parallel fences at a suitable distance to the windward side. The stationary obstacle reduces the velocity of the snow-laden air causing the burden to be dropped upon the far side of the fence before reaching the highway. On level roads crossing prairie land drifts were forestalled in the design of the road by raising the grade line above the surrounding ground surface so that the wind could blow the snow from the graveled surface.

A steam shovel is shown, in the accompanying illustration, opening the Snoqualmie Pass across the Cascade Mountains, in the State of Washington, considered an engineering feat in 1922.

1925~ADOPTION OF UNIFORM SIGNS

As the disconnected sections of improved roads were joined into continuous long-distance routes, travelers were bewildered by the motley array of direction and information signs encountered in the different States. In the neighborhood of their home towns motorists could find their way in spite of the crude or missing direction signs. On a long journey in unfamiliar localities, however, a lack of consistent information became a source of continual delay, uncertainty, and confusion. It was to bring order out of this chaos of signs and thus help to speed the tourist upon his way that the American Association of State Highway Officials suggested the plan for marking the main roads of the country with standardized information and direction signs. At the request of the Association, the Secretary of Agriculture, on March 2, 1925, appointed a Joint Board of State and Federal highway officials,"to undertake immediately the selection and designation of a comprehensive system of through interstate routes, and to devise a comprehensive and uniform scheme for designating such routes in such manner as to give them a conspicuous place among the highways of the country as roads of interstate and national significance."

1925 ADOPTION OF UNIFORM SIGNS

This Joint Board, after requesting the several State highway departments to select the routes within their respective borders which should be included in such a system, reviewed these recommendations at regional and national conferences, before a nationwide coordinated system throughout the 48 States was adopted. This network designated the numbered system of United States highways.

The numbers assigned to the routes by the Joint Board had a special significance. Even numbers were assigned to the east-west roads and odd numbers to the highways running in a north-south direction. The more important transcontinental routes were designated in multiples of 10 beginning at United States Route 10, south of the Canadian boundary, through United States Route 40, across the midsection of the country, to United States Route 90 traversing the southern States from coast to coast. The historic north-and-south road always of great importance which paralleled the Atlantic seaboard was numbered United States Route 1 and the numbers increased towards the west until the road along the Pacific Coast was enumerated United States Route 101. The uniform numbered marker was adopted, which is in use today, consisting of a shield, on the face of which is given the initials U.S., the number of the route, and the name of the State in black on a white base.

The Joint Board also agreed upon a standard design of rectangular direction signs with black letters superimposed upon a white background. The warning and danger signs which have added so much to the safety of our highways were standardized both as to shape and color.

The standard United States shield numbered markers are easily distinguishable from the State route markers emblazoned for example with a covered wagon for Nebraska.

1928~ THE INTER~AMERICAN HIGHWAY

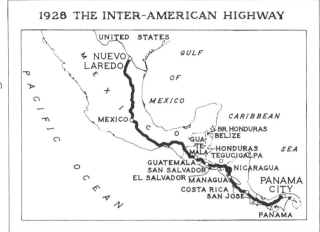

The Inter-American Highway, depicted on the map at right, constitutes the central section of the more extensive Pan American Highway proposed to join the North and South American republics. The subject of casual and fireside conversation long before it was a topic at the Pan American Highway Congress, assembled at Santiago, Chile, in 1923, cooperative arrangements to build the road were not initiated until 1928. On December 15, of that year, there was introduced in the United States Congress, and later signed by President Calvin Coolidge, a joint resolution authorizing a $50,000 appropriation to enable the Secretary of State to cooperate with the interested governments in making reconnaissance surveys to find the most practicable route for the proposed Inter-American Highway.

The Bureau of Public Roads, Department of Agriculture, was the agency selected to cooperate with the several governments, members of the Pan American Union, that signified a desire to participate in the surveys. Proceeding to Panama City, in June 1930, the Bureau's engineers established an office.

The reconnaissance surveys were speeded in the five participating countries for three years until 1933, when the field work was completed. The final report (Senate Document No. 224) was transmitted to President Franklin D. Roosevelt on March 5, 1934. The report concluded that the construction of the Inter-American Highway was feasible from an engineering standpoint. One-third of the road was estimated to have been completed by local authorities throughout the total 3,261-mile distance, later measured from Nuevo Laredo, on the United States-Mexico boundary, south to Panama City.

To begin actual construction operations on the Inter-American Highway, Congress appropriated $1,000,000, approved by President Franklin D. Roosevelt in June 1934. Another $75,000 was appropriated during the same month for making location surveys, plans and estimates. In March 1935, the Bureau of Public Roads opened construction headquarters in Panama City and later in San José, Costa Rica. The first work programmed consisted of bridging the three most important rivers, namely: the Tamazulapa, in Guatemala; the Choluteca, in Honduras; and the Chiriqui, in Panama.

With the preliminary arrangements established upon a firm footing only time was needed to interest the United States Government in more extensive cooperation. In December, 1936, at the Conference of the American Republics' meeting at Buenos Aires, Argentina, a Pan American Highway Convention was signed which declared for "the speedy completion of a Pan American Highway." The financing of the work was the crux of the problem. To relieve this situation the Export-Import Bank of Washington, on May 25, 1939, initiated arrangements to extend credit for public roads construction in Latin America.

Because of physical limitations, construction work dragged along after the outbreak of the war until the Japanese attack upon Pearl Harbor threatened to sever our coastwise communications with the Panama Canal. To complete an overland connection, President Franklin D. Roosevelt signed, on December 26, 1941, a $20,000,000 authorization of funds to be devoted to the building of the Inter-American Highway. Thereafter work was resumed in earnest and further accelerated in 1942 when the United States Army undertook the immediate construction of a usable pioneer trail for the entire length of the highway.

1933~ROADS TO SERVE THE LAND

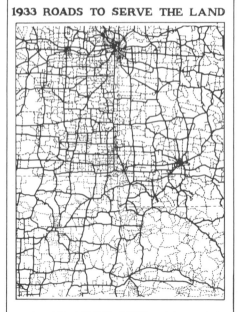

1933 ROADS TO SERVE THE LAND

To provide work for the unemployed, road improvement was quickened, in 1933, by large grants of funds made by the Federal Government. The country was in the depths of the financial depression which followed the rapid decline in prices of securities traded on the Wall Street Stock Exchange during the gloomy financial week of October 24 to 30, 1929. The end of the National fiscal year, on June 30, 1930, found the Government debt increased $600,000,000 to a total of 16 billion dollars with a deficit of $902,000,000. Financial conditions were growing steadily worse. On September 22,1931, the United States Steel Corporation announced a ten per cent wage reduction for its employees effective October 10. Similar action was taken soon by Bethlehem Steel Corporation, General Motors, and other organizations. On January 22,1932, the Reconstruction Finance Corporation bill was approved by President Herbert Hoover, providing two billion dollars for the emergency financing of distressed banks, building and loan societies, railroads and agriculture. This action arrested the decline only partially. On May 7, the United States Steel Corporation, bellwether of American industry, made a further wage reduction of approximately fifteen per cent. People in all walks of life were suffering deprivations. The bonus expeditionary army composed of unemployed persons, wives, children and little babies arrived in the National Capital on May 29, after gathering reinforcements all along their line of march. They came to demand from Congress help for the people out of work. They received an answer on June 21 when the Unemployment Relief Bill was approved providing two billion dollars, 90 per cent of which was to be lent to States and cities for relief work on self liquidating projects.

The new Administration, swept into office by the widespread resentment of the people, initiated vigorous measures to turn the tide of the depression. President Franklin Delano Roosevelt by executive order reduced Federal salaries, pensions and other benefits netting a saving of 400 million dollars. Congress passed the Industrial Recovery Bill authorizing $3,300,000,000 for public works on May 26. Some 275,000 unemployed men began to be mobilized on July 1, into a Civilian Conservation Corps distributed in camps throughout the country. On August 1, the National Recovery Act blanket code went into effect which prescribed maximum working hours and minimum wages for industry and trade which would guarantee a living to the employees.

The National Recovery Act assigned $400,000,000 to the States for highway construction realizing that roadwork was an effective method of bringing work right to the doors of the unemployed. Title II Section 204 assigned $94,676,687 for Class III secondary highway projects on secondary or feeder roads not now in the approved system of Federal-aid highways, but which are either part of the State highway systems or are important local highways leading to shipping points, or which will permit the coordination or extension of existing transportation facilities. This was the first major Federal action directed towards secondary or farm-to-market roads as a class.

1934—RAILROAD CROSSINGS BRIDGED

Ever since the total death toll from motor-vehicle accidents had passed the 30-thousand mark in 1929 efforts had been intensified by highway administrators and automobile manufacturers to make travel safer. Although representing only 5 per cent of the total motor-vehicle fatalities, the accidents at highway-railway grade crossings often resulted in such frightful catastrophes that the average citizen was shocked upon reading the harrowing details in the newspapers especially when a school bus of innocent children were the victims. Thus everytime that a serious grade-crossing accident occurred there was widespread public clamor for removing the cause by separating the grades or relocating the highway. Actually the elimination of these danger points one by one was the only practicable solution because the overall cost of doing away with all grade crossings would require funds reaching a total of astronomical proportions.

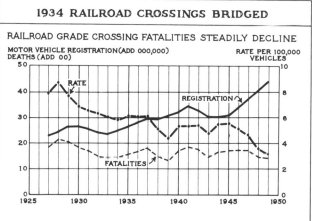

Highway engineers had been estimating the cost of this work for many years. For example on December 31, 1933, there were a total of 236,486 highway-railroad grade crossings on the 178,597 miles of Class I railways in the United States. Assuming an equivalent number of grade crossings per mile on all the railways of all classes in the country, totaling 243,857 miles, resulted in an estimated grand total of 320,600 grade crossings scattered from coast to coast. To eliminate all of these crossings by separation or relocation of the highway, estimated at $150,000 each, added up to the stupendous sum of more than 48 billion dollars. Obviously the best that could be done was to adopt a long-range program aimed at eliminating consistently year by year those grade crossings presenting the greatest hazard and causing the greatest delay.

There was a growing public concern with the whole subject of highway safety by 1934, because the total accident rate had increased, for the preceding two years, which followed the steadily falling curve showing the fatalities per 100 million vehicle miles from 1925 through the depression years to the low in 1932. For the first time in four years of adverse financial conditions the total motor vehicle registration, in 1934, showed an increase of 4.6 per cent over the preceding year. With more cars upon the road and the annual mileage per car rising, it was to be expected that the total number of accidents would increase. The number of fatalities caused by collision between motor vehicles and trains, however, remained about the same — 1,437 in 1933 and 1,457 in 1934. Public attention, however, was focused upon them because of their gruesome nature.

The importance attached to the elimination of highway-railroad grade crossings was evidenced in the National Industrial Recovery Act signed June 16, 1934, to provide work for the unemployed. Funds authorized by this law together with other Federal appropriations for the fiscal year ending June 30, 1934, aided in separating the grades of 70 intersecting highways and railways, part of a total of 405 grade-crossing-eliminating structures completed, under construction, or included in the program for removal during that year. Next the Emergency Relief Appropriation Act of April 8, 1935, allocated $200,000,000 for grade-crossing work. Since then the work has continued with encouraging results.

1935–VIADUCTS IN CITIES

The Emergency Relief Appropriation Act, approved by President Franklin D. Roosevelt on April 8, 1935, organized the first major attack upon the baffling problem of railroad grade crossings in cities. A great many rural, as well as a few metropolitan crossings had been eliminated in previous years with the aid of funds administered by the Bureau of Public Roads, United States Department of Agriculture. Just recently the $400,000,000 highway grant of the National Industrial Recovery Act, approved June 16, 1933, and the $200,000,000 highway grant of the Hayden-Cartwright Act, approved June 18, 1934, had laid great stress upon the need of ridding the roadways of the grade-crossing menace. Through the medium of this legislation Congress sought not only to provide work for the unemployed, but also to turn the rising tide of motor-vehicle accidents which had become a matter of growing concern to highway officials and road users. With the avowed purpose of promoting highway safety, these acts specified that the funds authorized should be devoted to "highway and bridge construction, including the elimination of hazards to highway traffic, such as the separation of grades at crossings, the reconstruction of existing railroad grade crossing structures, the relocation of highways to eliminate railroad crossings, ***** and the cost of any other construction that will provide safer traffic facilities or definitely eliminate existing hazards to pedestrian or vehicular traffic." Under the authority of these acts, as of June 30, 1935, on extensions of the Federal-aid highway system into and through municipalities there were completed 158 railroad-highway grade-separation structures, as well as 75 more placed under construction and others approved for construction.

Efforts to relieve unemployment by means of Federal highway appropriations had begun as early as 1930, following the financial crash of 1929. By 1933, the business depression had sunk to such a low level that Congress sought to apply emergency recovery measures aimed at critical areas of the national economy. The National Industrial Recovery Act and subsequent legislation authorized the expenditure of funds not only upon the approved Federal-aid system but also upon its extension into and through cities, where were concentrated large blocks of our population, and also upon feeder roads leading into the outlying rural areas where work was scarce. It was a happy coincidence that these extensions of the two extremes of the Federal-aid system occurred at a time when its initial improvement virtually had been completed. For more than a decade following the establishment of the Federal-aid highway system in 1921, the efforts of the State highway departments and the Federal authorities had been devoted to making the entire network equally serviceable for travel. In 1932, with the segments of this initial construction almost joined into a continuous network, the traffic obstacles shifted to the metropolitan congested areas. There also the accident risk was at a maximum because of the concentration of motor vehicles and pedestrians. It was logical, therefore, that viaducts across wide bands of parallel railroad tracks, represented in the accompanying illustration, should provide a promising project for the roadbuilder. By a single major operation the construction of an overpass of this type would do away with a prolific source of accidents as well as abolish the time-consuming delays to traffic caused by shifting freight trains.

1936~ROADSIDE BEAUTY RESTORED

Prominent among the objectives financed by the Public Works funds authorized in the National Recovery Act, approved June 16, 1933, and in subsequent legislation, was the landscaping of a moderate mileage of main roadsides, as shown in the accompanying illustration. The rules and regulations required that at least one-half of one per cent of each State's apportionment should be devoted to this type of improvement. Thus a total of approximately $2,000,000 was set aside for pioneering work which had for its ideal the conversion of unsightly roadsides into attractive areas bordering roadways made safe for travel and business. This broadening of the agenda of our road-building activities came at a psychological moment on the time clock of our now well-established highway program. Since 1921 the application of Federal-aid funds had been limited to a system of interstate and intercounty roads containing not more than 7 per cent of our total rural road mileage. Over the next decade and a half, Federal funds were restricted to the betterment of some 200,000 miles of main roads included in this system. By 1935 the initial improvement of these highways had reached the point where it was possible to travel from one end of the country to the other with at least a modicum of speed and comfort. Having accomplished the primary aim of "getting the traffic through" it followed as a logical sequence that attention should turn to putting on the finishing touches with such items as landscaping and planting of roadsides and the building of foot paths. These features embodied much more than an aesthetic value because debris-cluttered roadsides, the encroachment of unsightly advertising signs and shoulders overgrown with high weeds multiplied the accident risks at a time when the growing number of automobiles were straining at the seams of our threadbare garment of main highways.

The State of Massachusetts had sought to recover the charm of her old roadsides as early as 1912. In the late 1920's the Westchester County authorities in New York engineered the artificial restoration of scenic beauty to parkways on Long Island. In 1933, the Federal Government completed the Mount Vernon Memorial Highway leading from the National Capital to the home of George Washington. The superb landscaping of this historic thoroughfare included the selection of the most imposing trees, the opening of vistas along the Potomac River, the trimming or removal of unsightly vegetation, the planting of new trees and shrubs and the sowing of the intervening areas with grass seed or other ground cover. The wisdom of the original arrangement became apparent as the natural growth matured with the passing years. This project was a masterpiece of the cubical treatment of highway development which integrated the traveled way, the roadside and the adjoining landscape as elements in the final balanced highway unit. Space design involved much more than the removal of disorderly conditions from existing highways. The panorama of the spatial vision encompassed the acquirement of wide right of ways, the planning of the successive stages of improvement to serve the growing requirements of traffic, the rounding of shoulders and the provision of shallow ditches for safety and the merger of the roadside with the surrounding landscape so that the finished thoroughfare harmonized with the natural setting of the terrain.

1938~CITY ENTRANCES AND BELT LINES

With the initial surfacing virtually completed on the Federal-aid highway system, consisting of the principal State and interstate highways, there arose a widespread and justifiable demand for the extension of these main roads into and through centers of population by means of trans-city arteries and belt-line distribution thoroughfares, or ring roads, as shown in the accompanying illustration. Now that there were 29,442,705 motor vehicles registered, as compared with 1,258,060 twenty-five years before, the annual number of motor-vehicle fatalities had zoomed from 4,227 in 1913, to 32,400 in 1938, not to speak of the millions of persons injured and the vast financial losses resulting from property damage. Of the three primary elements of the accident problem, namely: the roadway, the vehicle and the driver, the attention of roadbuilding authorities naturally was concentrated upon their major responsibility—the provision of safer highways where there were the greatest concentrations of traffic.

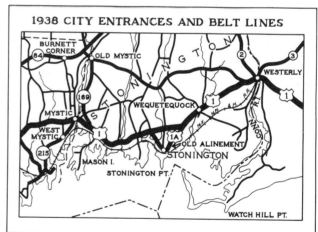

1938 CITY ENTRANCES AND BELT LINES

The Congressional sounding board had aroused intense public interest in a proposed system of multiple-lane superhighways built according to the highest standards of grade and alinement, with opposing streams of traffic separated by central parkways and grade-separation structures at all highway and railroad crossings, and access for side roads permitted only at carefully selected locations. The Federal-aid Highway Act of 1938 directed the Bureau of Public Roads, United States Department of Agriculture, to investigate the feasibility of building these superhighways.

The findings of the Bureau of Public Roads with respect to these superhighways were presented by President Franklin D. Roosevelt, in 1939, to the first session of the Seventy-sixth Congress in House Document No. 272, entitled "Toll Roads and Free Roads." The report concluded that the construction of six superhighways criss-crossing the country was practicable from the physical standpoint but unsound for financial reasons. The approximate total length of these six superhighways was estimated at 14,336 miles and the aggregate cost of building to desirable standards added up to $2,899,800,000—an average of $202,270 per mile. Spreading the cost of construction, maintenance and operation over the 15-year period, from 1945–60, the average expenditure was computed as $12,840 per mile per year financed by 30-year bonds bearing an interest rate of 2.6 per cent and an additional 2.24 per cent set aside each year as a sinking fund to retire the bonds at maturity. The total utilization of the six superhighways for the period of 1945–60 was estimated at 4,544,000,000 vehicle miles equivalent to an average daily traffic on each mile of the six superhighways of 699 passenger vehicles and 175 motor trucks. It was considered reasonable to collect not more than 3.5 cents toll per vehicle-mile for motor trucks and 1.0 cent for passenger cars. Upon this assumption the average annual toll collection over the period of 1945–60 was estimated to total $72,140,000, or less than half of the $184,054,000 average total annual cost of the six superhighways. It was concluded, therefore, that a direct toll system would not be productive of the funds required to recover the entire cost of these highway facilities. Having outlined the fallacy of the toll system, which had been proved again and again by the failures of toll roads in the past, the report presented a preferable alternative consisting of a master highway plan for the entire Nation embodying:

The construction of a system of interregional highways with necessary connections through and around cities.

The creation of a Federal Land Authority to acquire, hold and sell lands for highway right of ways.

1939—A HIGHWAY MASTERPIECE

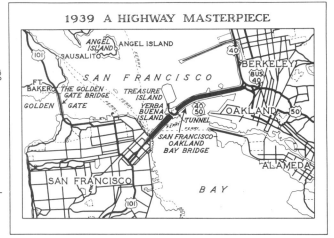

"An acetylene torch in the hands of Governor Frank F. Merriam burned asunder a heavy chain barrier; an electric button pressed by President Roosevelt in the White House in Washington flashed the green 'Go' signal and three columns of whirring automobiles sped from each shore of San Francisco Bay over six lanes of the world's greatest aerial highway — the San Francisco-Oakland Bay Bridge — a half hour after noon on November 12,1936." Thus began an article describing the opening in the California Highways and Public Works Magazine.

For nearly a century San Francisco's leaders had dreamed of a great bridge speeding communications with the East Bay Empire served by the prosperous cities of Oakland, Berkeley and Alameda, wrote C. H. Purcell, the chief engineer in charge of the construction of this masterpiece of engineering genius. Back in 1850, William Walker, a militant San Francisco newspaper editor, proposed a causeway from the Golden Gate city to Contra Costa County patterned after the well-known 2,000-foot Clay Street wharf with foundations as deep as forty feet. Thereafter interest in the project lagged until revived in 1856 by William Tecumseh Sherman, a young San Francisco banker, later a celebrated Civil War general. No more appealing proposals were made until 1869, when the Central and Union Pacific Railroads were joined into a single transcontinental unit. At that time Leland Stanford, later United States Senator from California, urged upon his railway the desirability of a railroad bay crossing.

More than half a century elapsed before serious attention was directed again at the San Francisco-Oakland Bay Bridge Project. In 1921, the San Francisco Car Dealers Association financed an engineering study to determine the feasibility of a combined tube and concrete causeway crossing. Within seven years 35 proposals to build such a connection had been submitted by corporate and individual bidders. San Francisco's Mayor James Rolfe, Jr., in 1928, headed a delegation urging Congress to pass a bill authorizing a bay bridge. The legislation was defeated by Army and Navy objections relating to national defense and interference with navigation. Arriving at the conclusion that Californians must bear the burden if the bridge were to be built at all, a Toll Bridge Authority was established by the California State Legislature in 1929. Congressional approval of a $73,000,000 loan from the Federal Reconstruction Finance Corporation was obtained in June, 1932, with the endorsement of President Herbert Hoover. Ground breaking began for the bridge on July 9, 1933, and forty months later the completed structure was opened to automobile and truck traffic.

Before the gigantic crossing was built thirty-five million ferry-boat commuters annually passed over the bay.

During the first year of operation 25,000 vehicles daily traversed the San Francisco-Oakland Bay Bridge, shown in the accompanying illustration. After the competing ferries halved their fares in 1937, the bridge traffic slumped temporarily to 23,600 vehicles per day. When the Golden Gate Exposition opened in 1939, the bridge automobile tolls were reduced from 50 to 40 cents. With this stimulus the bridge traffic rose to 30,000 daily vehicles. During 1940, when tolls were cut further to 25 cents, the ferries went out of business.

1941—NATIONAL DEFENSE ROADS

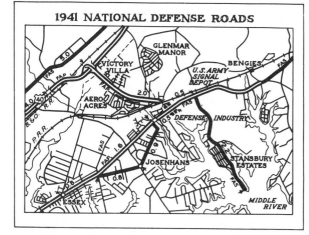

Spurred into action by the grave turn in the European fortunes of war, which so menaced the survival of our historic allies in the Old World that the future security of our own country seemed to be at stake, Congress passed the Federal Highway Act, approved by President Franklin D. Roosevelt on September 5, 1940. This legislation authorized the Commissioner of Public Roads to "give priority of approval to, and expedite the construction of, projects that are recommended by the appropriate Federal defense agency as important to the national defense."

The planning of a strategic highway system had its origin shortly after the conclusion of World War I. After the Federal-aid highway system was established by the Federal Highway Act of 1921, the Bureau of Public Roads, Department of Agriculture, sought the advice of the War Department concerning the location and character of highways needed for national defense. The outcome of these conversations was the Pershing map of 1922 on which were indicated for the first time the main roads selected by responsible military authorities as of prime importance in time of war. These routes in the ensuing years were improved with Federal and State funds. From time to time the Pershing map was revised with the advice of the War and Navy Departments. On May 15, 1941, within a year after the passage of the Federal Highway Act of 1940, a revision was made.

At this date, because of our national policy of neutrality, no special funds had been authorized by Congress in preparation for war. The planning and construction of national defense roads, however, was already a fixed policy to the maximum possible extent permitted by the limitations of Federal-aid road legislation. In May, 1941, a Joint Action Board had been formed consisting of representatives of the Corps of Engineers, the Transportation Corps and the Services of Supply, War Department, the Bureau of Yards and Docks, Navy Department, and the Public Roads Administration.

The need for more money to finance our constantly expanding war activity was supplied by the $150,000,000 fund authorized in the Defense Highway Act, approved on November 19, 1941. The act legalized the construction of access roads to military and naval reservations and to sites of defense industries and sources of raw materials when certified as important to national defense by the Secretary of War or of the Navy. The act authorized also a sum of $10,000,000 for the construction of aircraft flight strips to be used as auxiliary landing areas at strategic locations, or for intermediate stations on long flights involving supplies, equipment and troops, or as intercepting bases to ward off possible enemy bomber attacks, or as operational locations for coastal patrol. The act further provided $25,000,000 for apportionment among the States for correction of critical deficiencies in the strategic network, and an additional $25,000,000 for the same purpose to be employed without regard to apportionment. Such was the status of national defense work on that grim morning of December 7, 1941, when the Nation awoke to hear the stunning news of the Japanese surprise attack on Pearl Harbor.

Thereafter approval of highway projects was restricted to those roads essential to implementing our declaration of war. Vigorous efforts were devoted to the removal of traffic bottlenecks in the vicinity of military camps, munitions plants and shipyards. Simultaneously the Public Roads Administration, in cooperation with the Office of Price Administration and the War Production Board, took steps to defer all non-critical highway work.

1943~THE ALASKA HIGHWAY

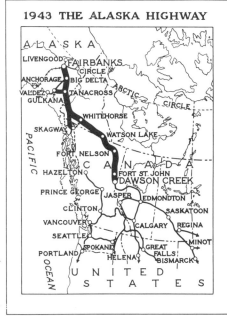

Talked about as early as 1910, the Alaska Highway crystallized from camp-fire conversation into the bold outlines of reality when an International Committee was appointed in 1930, by joint agreement of the governments of the United States and Canada, to study the feasibility of a public road leading to our northern Territory. The American members of the committee submitted their report to President Franklin D. Roosevelt on May 1, 1933. Thereafter action was delayed by the more pressing demands for road construction in populous areas to relieve unemployment during the protracted financial depression. Governmental interest in the project, revived in 1938, culminated in a favorable report made in 1940 by a committee of Congress. Then, in 1941, the devastating Japanese attack upon Pearl Harbor alerted our military authorities to the defenseless character of our Pacific coast shipping lanes. Failure to defend the Aleutian Islands and Alaska might endanger the entire west coast to landings of amphibious enemy forces. Overnight Alaska was transformed from the status of a secondary defense area to a region of primary strategic importance not only for immediate defense but also as an eventual base for aerial attack upon the Japanese homeland. Faced with these stern realities, the War Department General Staff concluded that Alaska must be protected by war planes. To implement this decision, they recommended the rapid construction of a highway joining the chain of airports linking Edmonton, Alberta, with Fairbanks, Alaska. This road was to be used for transporting supplies, men and equipment to the strategic airfields and to provide safety for fliers engaged in ferrying aircraft from the United States to Alaska and Russia.

Motivated by the dictates of military necessity, the administrative steps preliminary to the building of the Alaska Highway followed one another with the steady rhythm of machine-gun fire.

Preparations for the biggest single highway construction job on record were speeded while the governmental negotiations were in progress. Early in March, the Public Roads Administration began planning to move engineers into the frozen northern wilderness by airplanes, on snowshoes and with Husky dog teams to explore the mountain passes for the most practicable route. Engineer troops were rendezvousing at the Dawson Creek railhead by March 10. During the working season of 1942, seven regiments of 10,000 work troops, wearing arctic clothing for protection against sub-zero temperatures, cut the pioneer road through the forests and across the frozen muskeg, with the aid of bulldozers, for the entire distance to Fairbanks. Meanwhile the Public Roads Administration established a district office at Seattle, Washington, and field offices at White Horse and St. John as bases for four management contractors who had assembled 47 construction contractors. During the 1942 season 500 Public Roads Administration engineers and aids directed the work of 7,000 contractors' roadbuilders who rushed the clearing, grading, draining, timber-bridge construction and gravel surfacing of a 20- to 24-foot roadway. In the following year, on October 31, 1943, the Public Roads Administration, with the aid of 81 contractors employing 14,000 civilian workmen, using 6,000 heavy units of roadbuilding machinery, turned over the completed road to the United States Army. Thus, in about twenty months, a 1,480-mile motor-vehicle road was opened to travel through a virgin wilderness.

1945–A RURAL INTERSTATE HIGHWAY

The National Interregional Highway Committee was appointed by President Franklin D. Roosevelt on April 14,1941, "to investigate the need for a limited system of national highways to improve the facilities now available for interregional transportation, and to advise the Federal Works Administrator as to the desirable character of such improvement, and the possibility of utilizing some of the manpower and industrial capacity expected to be available at the end of the war."

"The committee, with the aid of a staff provided by the Public Roads Administration, made careful and extended studies of the subject" and submitted to the President a report which was transmitted to the Seventy-eighth Congress, Second Session, as House Document No. 379. The report recommended "the designation and improvement to high standards of a national system of rural and urban highways totaling approximately 34,000 miles and interconnecting the principal geographic regions of the country. *****"

The President continued, "The improvement of a limited mileage of the most heavily traveled highways obviously represents a major segment of the road replacement and modernization program which will confront the Nation in post-war years.

"Continued development of the vast network of rural secondary roads and city thoroughfares, which serve as feeder lines and provide land-access service, likewise has an important place in the over-all program, together with the repair or reconstruction of a large mileage of Federal and State primary highways."

In the report accompanying the President's message, the committee concluded that preferential right of way should be accorded traffic moving over the interregional routes selected to be the main collectors of the national travel. The speeding of express traffic with due regard to safety and economy required the reduction to a minimum of the access roads and crossings by limiting the entrance upon the main routes to carefully chosen points.

The Committee recognized that many unimportant rural cross roads could be closed and their traffic diverted to more convenient crossings. Again where the travel upon the interregional highway was light, it might not be necessary to build at once a grade-separation structure. Wherever, however, a grade crossing was permitted, as shown in the accompanying illustration, the intersection was to be so designed and signed that the presence of vehicles upon the main highway would be apparent to entering vehicles. Furthermore, all traffic was to be halted at grade crossings.

The recommendations of President Roosevelt's interregional highway study as well as of his 1939 report entitled, "Toll Roads and Free Roads" were incorporated in the Federal Aid Highway Act, approved December 20, 1944. This legislation authorized a Federal appropriation of $500,000,000 for each of the first three post-war years. Of this total amount $225,000,000 was to be spent upon the Federal-aid system in rural regions and $125,000,000 in urban areas. For the construction of secondary and feeder roads $150,000,000 was designated. The act provided for the selection of a National System of Interstate Highways not exceeding 40,000 miles in extent. In February, 1945 the Public Roads Administration requested each State Highway Department to proceed at once with recommendations of routes for inclusion in the system without any limitations upon their freedom of action.

1945—URBAN DEPRESSED EXPRESS HIGHWAY

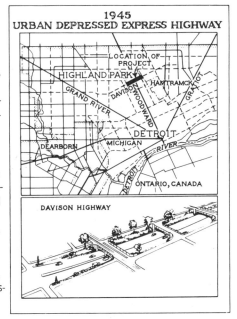

The report pertaining to Interregional Highways, transmitted to the Congress on January 12, 1944, contained this significant statement, "The city streets over which the urban mileage included in the recommended interregional system has been measured, are those now marked as the transcity connections of the existing main rural highways that conform closely to the rural sections of the recommended routes. These streets generally pass through or very close to the existing central business areas of the cities."

The Committee which signed the report observed that, "the studies made of 3 cities of 300,000 or more population show that upward of 90 percent of the traffic moving toward these cities on main approach highways consisted of vehicles bound to ultimate or intermediate destinations within the cities themselves. For the 4 cities of 50,000 to 300,000 population, the similar proportion of city-bound traffic was found to be above 80 percent. For smaller cities, the corresponding proportion tends to decline, reaching 50 percent for the cities of less than 2,500 population that were studied."

Since the transcity connections joining the rural portions of the Interregional Highway System must penetrate through or near the heart of the city, the problem resolved itself into how best to reduce the number of street intersections which were prolific sources of delays as well as accidents. There arose also the problem of preventing the gradual choking of the main thoroughfare by the progressive encroachment of ribbon developments. Studies made by the Public Roads Administration showed that a one-way two-lane highway with no intersections could discharge without undue congestion 3,000 vehicles per hour at an average speed of 35 miles per hour. With three traffic lights per mile, each set on a half-minute interval, the hourly discharge was cut in half to 1500 vehicles.

The Committee recommended, therefore, in the largest cities that the interregional routes, in order to avoid intersections at grade, be raised above or depressed below the natural ground level. Turning from elevated routes because of their tendency to divide the city and depreciate abutting property, the Committee discussed the pros and cons of depressed expressways. They held that this type of improvement was pleasing to the eye and in keeping with the urban environment. The depressed route, however required extensive reconstruction of underground facilities including watermains, sewers and electric conduits. There was the added difficulty of drainage necessitating the installation of mechanical pumps where the terrain was level.

Neither the elevation nor the continuous depression of expressways was recommended in the outer and residential sections of large cities nor generally in small cities. Under such circumstances the design, shown in the accompanying illustration, was preferred. This parklike development may be built within a block-wide right-of-way employing long, rolling grades to pass under bridges constructed at the cross-street levels, with access connections and pedestrian overpasses at appropriate intervals. In this design the existing surface streets are to be utilized as lateral service ways.

The graceful curves of two-leveled State Route 17 preserve the quiet beauty of the low valleys of the Catskill Mountains. *New York State Department of Transportation*

EPILOGUE: Today's Highways

This history of America's highways by Albert C. Rose and Carl Rakeman ends with World War II and the immediate postwar era. But the story of America's *modern* highways begins where Rose and Rakeman end. Therefore, our modern highway story begins with the super highways—the Interstate System.

The Interstate System, technically known as the National System of Interstate and Defense Highways, includes the most important highway routes of the nation. The need for such a system was first described by the Bureau of Public Roads in a report to Congress in 1939 and was further justified in follow-up studies. Then, in 1944, Congress, acting on the recommendations, directed that an Interstate System be created. In the Federal-Aid Highway Act of 1944 it provided that ''there shall be designated within the continental United States a National System of Interstate Highways . . . located as to connect by routes as directly as practicable the principal metropolitan areas, cities, and industrial centers, to serve the national defense, and to connect at suitable border points with routes of continental importance in the Dominion of Canada and the Republic of Mexico.''

In compliance with the act, there was designated, on August 2, 1947, an Interstate System of highways—selected through joint action of the states and the Bureau of Public Roads—totaling an estimated 37,700 miles in length.

Subsequent System adjustments, additions, and deletions have resulted in the present total of 42,500 miles authorized by Congress for the Interstate System.

Crisscrossing the nation with 86 percent of its miles open to traffic in 1975, the Interstate System links more than 90 percent of our cities with populations of 50,000 or more, plus many smaller cities and towns. It will serve well over half the urban—and almost half the rural—population of the country. When it is completed, the Interstate System, comprising little more than 1 percent of the nation's total road and street mileage, will carry 20 percent of all the traffic in the United States.

Interstate projects are planned so as to accommodate adequately traffic anticipated twenty years hence. All routes are at least four-lane divided highways, becoming six and eight lanes in and near the large metropolitan areas.

Each traffic lane is twelve feet wide. Median areas between roadways of a divided highway are generally at least thirty-six feet wide; lesser widths are used where necessary in mountainous and urban locations. Wide right-of-way is needed to provide for medians and shoulders as well as pavements. Aesthetic design must adapt the road to the terrain, such as placing two roadways on opposite sides of a stream, or at different levels on a hillside. Free-flowing design with variable-width medians often saves money, and at the same time avoids the monotony that can lead to inattentive and dangerous driving. Traffic interchanges along the System provide frequent and safe access, and overpasses and underpasses eliminate all highway and railroad grade crossings. (Where needed, in developed areas, frontage roads for local traffic are provided on one or both sides.) It will be possible to drive from coast to coast or border to border without encountering a traffic light or stop sign.

And although federal law does not permit commercial facilities within the Interstate right-of-way, signs advise a motorist when he is approaching connecting roads that lead to nearby gas stations, restaurants, and motels.

The need for the Interstate System is obvious. We pay dearly for inadequate roads and streets; not just in deaths, injuries, damages, frazzled nerves, inconvenience, time, and gasoline but in the price of everything we buy or sell. Studies show that highway-user benefits of the Interstate System will far exceed the cost when the System is completed. These benefits stem from lower operating, time, accident, and impedance (strain of driving) costs. Standards for the Interstate System, planned jointly by the states and the Federal Highway Administration, incorporate the latest proven features that provide for safe and tension-free driving, for economy of vehicle operations, and for pleasing appearance. Sweeping curves, easy grades, and long sight distances provide for safe driving at reasonable speed. Actual speed regulations, as on all state highways, is under the control of the states.

Overall, the System provides new freedom and new speed and safety to the movement of people and goods (travel on the Interstate is more than twice as safe as on older roads). The advantages to long-range travel, whether by car, bus, or truck, are obvious: Business and vacation travel take less time than formerly. There is more comfort and less strain in driving. Deliveries are faster and trucking operations more efficient. Interstate routes are important parts of the production, assembly, and distribution lines of business and industry. Farm products are shipped more quickly and with less loss in spoilage and quality.

By going into and through our large cities and skirting the central business districts, the System helps eliminate today's traffic jams. It speeds commuters and shoppers from the suburbs—linked by circumferential or belt routes that separate the suburbs from traffic headed into and out of the city centers. The System also bypasses smaller cities and towns, providing access to them but keeping through-traffic off the congested main streets, freeing them to handle local traffic.

The construction of the Interstate System has created many jobs and additional markets. There are new jobs in roadbuilding, plus increases in production of such construction equipment as steel, cement, bituminous materials, aggregates, and so on.

Dramatic growth and development of business along the Interstate System has occurred. However, traffic from the businesses, industries, and homes that are springing up alongside the System enters the main stream only at interchanges, helping to prevent slowdowns or congestion.

Interstate 95, between Kittery, Maine, and Portsmouth, New Hampshire, featuring the light, graceful 1,344-foot steel structure of the Piscataqua River bridge. *New Hampshire Department of Public Works and Highways; Maine Department of Transportation*

Interstate 64 disappears into twin tunnels, thus preserving historic Cochran Hill in Louisville's Cherokee Park. *Kentucky Department of Highways*

Building the Interstate, unlike constructing a single new road, is a complicated business. State, county, and city lines are crossed, and all the governments, as well as the federal government, are concerned with the route locations and their effects. About 85 percent of the System is being built on new locations.

Over 2,300 miles of toll roads, tunnels, and bridges have been included in the System, as permitted by law. They were built without federal aid. (Federal law prohibits charging tolls on any road section improved with federal-aid funds).

Before a shovelful of dirt could be moved in the construction of Interstate routes, future traffic and its economic and social effects had to be forecast. Detailed locations had to be selected; the ideas and plans of cities and counties had to be considered and coordinated; surveys and plans had to be made; bridges and interchanges designed; and right-of-way and access control acquired.

The Federal Highway Administration is required to provide the Congress every few years with an estimate of what the entire cost of the Interstate System will be when completed. The 1975 cost estimate placed this figure at $89.2 billion, of which the federal share was $79.52 billion.

The present estimate of the cost of completing the Interstate System is substantially greater than the original estimates. Several factors have contributed to this. Soaring inflation, of course, is a major factor. Also, the System design has been changed to improve road capacity; additional safety and durability features are being incorporated in System design as a result of Congressional direction or operation experience; and additional funds are being spent to protect the environment or beautify the highway rights-of-way.

The law provides that the amount of federal aid authorized each year for the Interstate program be distributed among the states according to a long-established formula.

Construction of the System is nearing completion in practically all the rural areas and in most urban areas. To summarize the benefits to the nation, the movement of people and goods on this System will result in a user-benefit ratio of about $2.90 for every dollar invested in construction over its service life, the differential in operating safety over other highway systems will add up to thousands of additional lives saved annually, the System is a dramatic contribution to the economic development of the areas it serves, and, because of its advanced design features, it provides continued service long after other highways would have become obsolete.

How to integrate the Interstate System into the nation's communities and culture is a complex problem but one that is critical to our sociological, environmental, and economic well being. The photographs included here are examples of highways designed for safety and beauty as well as for ecological and engineering considerations.

Norbert T. Tiemann
Administrator, Federal Highway Administration

Washington, D.C.
January, 1976

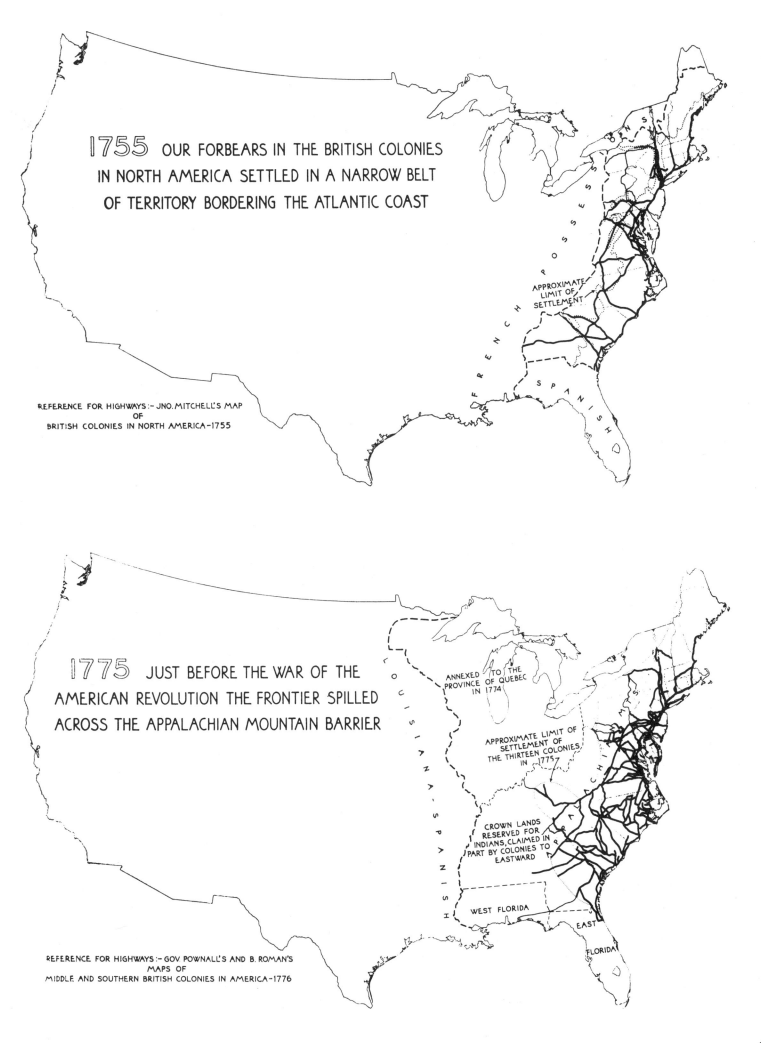

1755 OUR FORBEARS IN THE BRITISH COLONIES IN NORTH AMERICA SETTLED IN A NARROW BELT OF TERRITORY BORDERING THE ATLANTIC COAST

FRENCH POSSESSIONS

APPROXIMATE LIMIT OF SETTLEMENT

SPANISH

REFERENCE FOR HIGHWAYS:- JNO. MITCHELL'S MAP OF BRITISH COLONIES IN NORTH AMERICA-1755

1775 JUST BEFORE THE WAR OF THE AMERICAN REVOLUTION THE FRONTIER SPILLED ACROSS THE APPALACHIAN MOUNTAIN BARRIER

LOUISIANA-SPANISH

ANNEXED TO THE PROVINCE OF QUEBEC IN 1774

APPROXIMATE LIMIT OF SETTLEMENT OF THE THIRTEEN COLONIES IN 1775

APPALACHIAN MTS.

CROWN LANDS RESERVED FOR INDIANS, CLAIMED IN PART BY COLONIES TO EASTWARD

WEST FLORIDA

EAST FLORIDA

REFERENCE FOR HIGHWAYS:- GOV. POWNALL'S AND B. ROMAN'S MAPS OF MIDDLE AND SOUTHERN BRITISH COLONIES IN AMERICA-1776

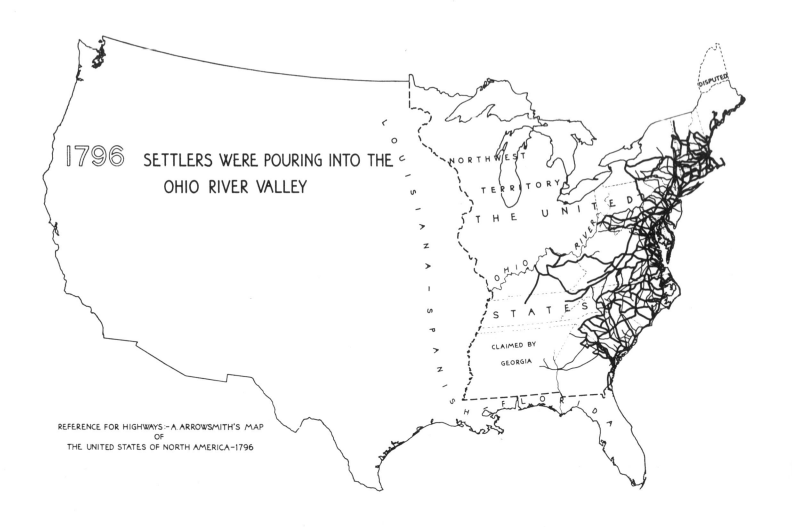

1796   SETTLERS WERE POURING INTO THE
OHIO RIVER VALLEY

REFERENCE FOR HIGHWAYS:-A. ARROWSMITH'S MAP
OF
THE UNITED STATES OF NORTH AMERICA-1796

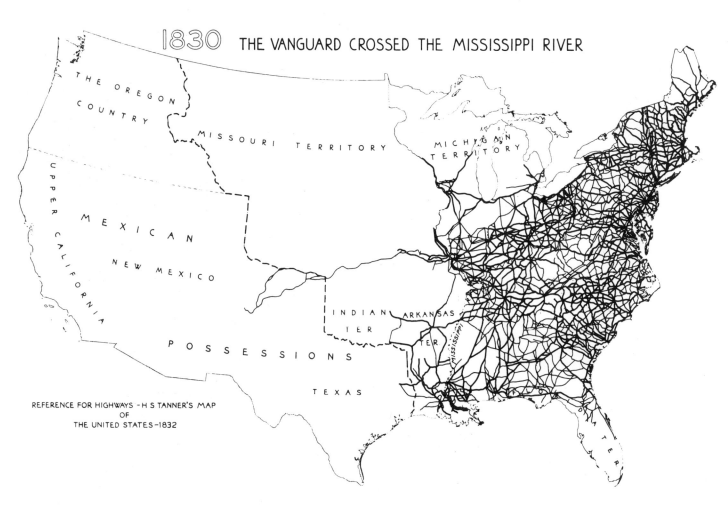

1830   THE VANGUARD CROSSED THE MISSISSIPPI RIVER

REFERENCE FOR HIGHWAYS - H S TANNER'S MAP
OF
THE UNITED STATES-1832

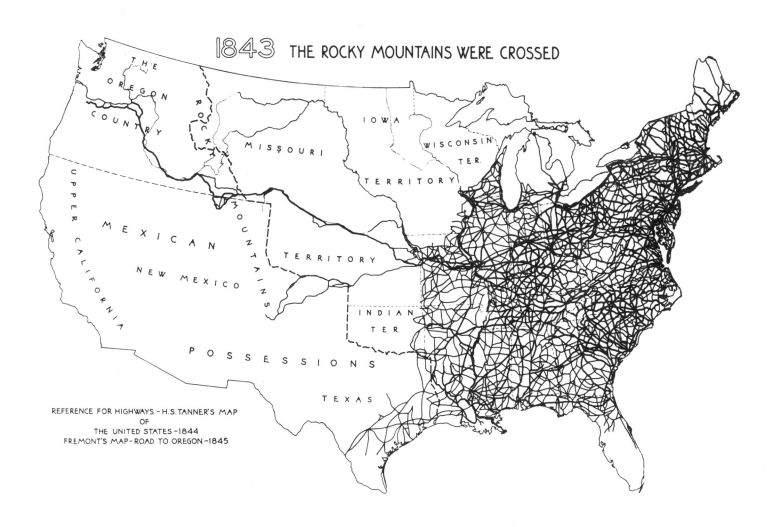

1843 THE ROCKY MOUNTAINS WERE CROSSED

THE OREGON COUNTRY

ROCKY MOUNTAINS

IOWA

WISCONSIN TER.

MISSOURI

TERRITORY

UPPER CALIFORNIA

MEXICAN

NEW MEXICO

TERRITORY

INDIAN TER.

POSSESSIONS

TEXAS

REFERENCE FOR HIGHWAYS:-H.S.TANNER'S MAP
OF
THE UNITED STATES-1844
FREMONT'S MAP-ROAD TO OREGON-1845

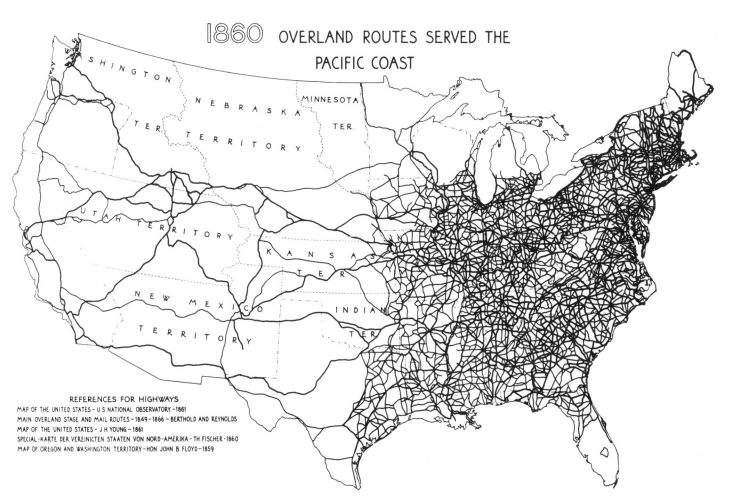

1860 OVERLAND ROUTES SERVED THE
PACIFIC COAST

WASHINGTON

NEBRASKA

MINNESOTA TER.

TERRITORY

UTAH TERRITORY

KANSAS TER.

NEW MEXICO

TERRITORY

INDIAN TER.

REFERENCES FOR HIGHWAYS
MAP OF THE UNITED STATES - U S NATIONAL OBSERVATORY -1861
MAIN OVERLAND STAGE AND MAIL ROUTES -1849-1866 - BERTHOLD AND REYNOLDS
MAP OF THE UNITED STATES - J H YOUNG - 1861
SPECIAL-KARTE DER VEREINICTEN STAATEN VON NORD-AMERIKA - TH.FISCHER-1860
MAP OF OREGON AND WASHINGTON TERRITORY -HON JOHN B. FLOYD -1859

TODAY THE COUNTRY IS CRISS-CROSSED BY THE BEST NETWORK OF PRIMARY HIGHWAYS EVER POSSESSED BY ANY NATION OF ANCIENT OR MODERN TIMES

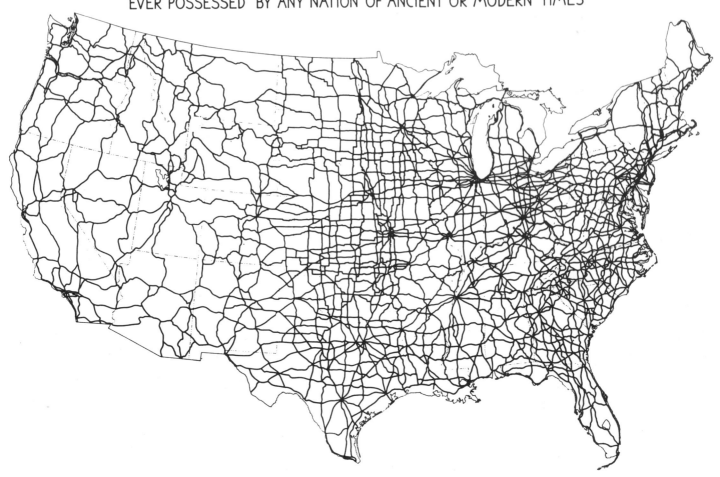

THE NATIONAL SYSTEM OF INTERSTATE AND DEFENSE HIGHWAYS

INDEX